公安院校招录培养体制改革试点专业系列教材

网络编程与动态网站制作

唐红杰 等 编著

清华大学出版社
北 京

内 容 简 介

本书精选网络编程与动态网站制作的核心内容，运用简洁的语言和丰富的图示，对抽象的理论知识进行了生动细致的描述。全书以JSP技术为主线，强调知识够用、突出能力培养，采用“教学做练”一体化方式撰写，合理组织教学单元，并将每个教学单元分解为知识要点、技能操作、拓展训练三个模块。内容共分为12章，具体包括网络编程与动态网站制作基础、搭建JSP动态网站开发环境、JSP基本语法、JSP内置对象、JavaBean的使用、JSP对数据库的访问、JSP与Servlet、基于Servlet的MVC模式、开发Web应用过滤器、表达式语言、标准标签库、动态网站开发综合实例等。书中实例侧重实用性、启发性和趣味性，深入浅出，通俗易懂，使读者能够快速掌握网络编程与动态网站制作的必备知识及编程技巧，为后续实战应用打下坚实基础。

本书知识与技能相辅相成、理论与实践有机融合，适合作为高等院校计算机应用、网络工程、软件工程等相关专业学生的授课教材，也适合作为动态网站设计与制作的培训教材，还可以作为动态网站设计与制作爱好者的自学教材。

本书封面贴有清华大学出版社防伪标签，无标签者不得销售。
版权所有，侵权必究。侵权举报电话：010-62782989　13701121933

图书在版编目(CIP)数据

网络编程与动态网站制作/唐红杰等编著. —北京：清华大学出版社，2016
公安院校招录培养体制改革试点专业系列教材
ISBN 978-7-302-41992-1

Ⅰ. ①网…　Ⅱ. ①唐…　Ⅲ. ①计算机网络－程序设计－高等学校－教材 ②网站－设计－高等学校－教材　Ⅳ. ①TP393

中国版本图书馆CIP数据核字(2015)第263186号

责任编辑：闫红梅　李　晔
封面设计：常雪影
责任校对：梁　毅
责任印制：何　芊

出版发行：清华大学出版社
网　　址：http://www.tup.com.cn，http://www.wqbook.com
地　　址：北京清华大学学研大厦A座　　**邮　　编**：100084
社 总 机：010-62770175　　**邮　　购**：010-62786544
投稿与读者服务：010-62776969，c-service@tup.tsinghua.edu.cn
质 量 反 馈：010-62772015，zhiliang@tup.tsinghua.edu.cn
课 件 下 载：http://www.tup.com.cn,010-62795954
印 装 者：北京密云胶印厂
经　　销：全国新华书店
开　　本：185mm×230mm　　**印　　张**：19.75　　**字　　数**：440千字
版　　次：2016年4月第1版　　**印　　次**：2016年4月第1次印刷
印　　数：1～2000
定　　价：39.50元

产品编号：064492-01

前言

本教材按照教、学、做、练一体化模式精编了网络编程与动态网站制作的核心内容，以知识要点、技能操作、拓展训练为单元组织教材的体系结构。知识要点给出了最重要和最实用的知识，是教师需要重点讲授的部分；技能操作给出了学生与教师共同完成的操作任务；拓展训练给出了需要学生独立完成的实践活动。

全书共分为12章。第1章主要介绍网络编程与动态网站制作的基础知识，重点讲述客户端开发技术 HTML 超文本标记语言、CSS 层叠样式表、JavaScript 脚本语言。第2章主要介绍服务器端开发技术 JSP 的初步知识，包括 JSP 动态网站开发环境的搭建及利用文本编辑器、IDE 集成开发环境开发简单 Web 应用的基本步骤。第3章介绍 JSP 基本语法，包括 JSP 文件的构成、成员变量和方法的定义、Java 动态元素以及常用的 JSP 指令标记和 JSP 动作标记。第4章介绍常用的 JSP 内置对象，包括 request、response、session、application 及其常用方法。第5章介绍 JavaBean 的使用，包括编写和使用 JavaBean、获取和设置 bean 属性等。JSP 和 JavaBean 技术的结合不仅可以实现数据表示和数据处理分离，而且还可以提高 JSP 程序代码的重用度，是 JSP 编程中常用的技术。第6章介绍 JSP 如何通过 JDBC-ODBC 桥接方式和纯 Java 数据库驱动程序方式实现对数据库的访问，如 Microsoft Access、SQL Server、MySQL、Oracle 等数据库。第7章和第8章讲述 Servlet 的运行原理以及基于 Servlet 的 MVC 模式应用。第9章介绍 Web 应用过滤器的基本概念、运行原理及其实际应用。第10章介绍表达式语言 EL 的基本用法，包括使用 EL 访问对象的属性、EL 内置对象等。第11章介绍标准标签库 JSTL 的基本用法，包括普通标签、条件控制标签、迭代标签等的使用。第12章通过一个动态网站开发综合实例讲解如何采用 JSP+JavaBean+Servlet 来开发 Web 应用项目。另有附录 A 为 HTML 标签库，附录 B 为 HTML 颜色库，附录 C 为 JSP 内置对象及其常用方法。

本书内容翔实丰富，注重实践操作，突出能力培养，适合作为高等院校计算机相关专业动态网站开发类课程的使用教材。书中第10章和第11章分别由辽宁警察学院的张爽老师和闫薇老师编写，其余章节由辽宁警察学院唐红杰老师编写并对全书进行统稿。另外，辽宁警察学院田静老师对书中图文进行了整理和校对。由于时间和作者水平有限，书中难免有欠妥之处，恳请读者批评指正。

编　者

2015年8月

前言

目 录

第 1 章

网络编程与动态网站制作基础

网络编程与动态网站制作通常包含客户端开发和服务器端开发两部分：客户端开发主要用于描述信息的内容，也就是浏览器中显示的网页；服务器端开发主要用于实现信息的处理，也就是前后台的动态交互。常用的客户端开发技术有 HTML、CSS、脚本语言等，常用的服务器端开发技术有 ASP、JSP、PHP 等。本章主要介绍客户端开发技术，后续章节将详细介绍服务器端主流开发技术 JSP 的相关内容。

1.1 HTML

1.1.1 知识要点

HTML(Hyper Text Markup Language，超文本标记语言)是一种用于描述网页的语言，它不是编程语言，而是标记语言。换言之，HTML 使用标记标签来描述网页，它不需要编译，可以直接由浏览器执行。HTML 标签是由尖括号包围起来的关键词，例如<html>；标签通常是成对出现的，例如<body>和</body>；标签对中的第一个标签是开始标签，第二个标签是结束标签，也有个别标签独立出现，如
和<p>；HTML 标签可以有一个或多个属性，属性的主要用途是辅助 HTML 标签来更美观地、更准确地描述网页的内容，例如<body bgcolor="cyan">中的 bgcolor 属性就描述了网页的背景颜色为青绿色。

HTML 标签和纯文本共同构成 HTML 文档，即通常所称的静态网页。客户端的浏览器能够读取 HTML 文档，并以网页的形式显示出它们。浏览器不会显示 HTML 标签，而是使用标签来解释页面的内容。

本书附录 A 给出了常用的 HTML 标签，附录 B 给出了常用的 HTML 颜色库，供参考。

1.1.2 技能操作

运用 HTML 语言制作简单的网页，具体任务如下：

编写一个网页，要求网页的标题为“公安网站示例”，网页背景是灰色，解释执行后在浏览器中显示两行文字——“两人被困事故车中 当地消防急速营救”和“组织党员干部开展反腐倡廉警示教育”，并且显示一张图片。具体的网页效果如图 1-1 所示。

完成任务的步骤如下：

- 编写、保存 HTML 文件。

- 使用浏览器打开网页。

1. 使用文本编辑器编写、保存 HTML 文件

首先需要打开一个文本编辑器。如果是 Windows 操作系统，可以通过“开始”→“程序”→“附件”→“记事本”命令来直接打开一个文本编辑器；如果是其他操作系统，可以参考操作系统的帮助手册打开一个文本编辑器。

图 1-1　HTML 网页示例

1）编写 HTML 文件的内容

在打开的文本编辑器中输入下面的内容：

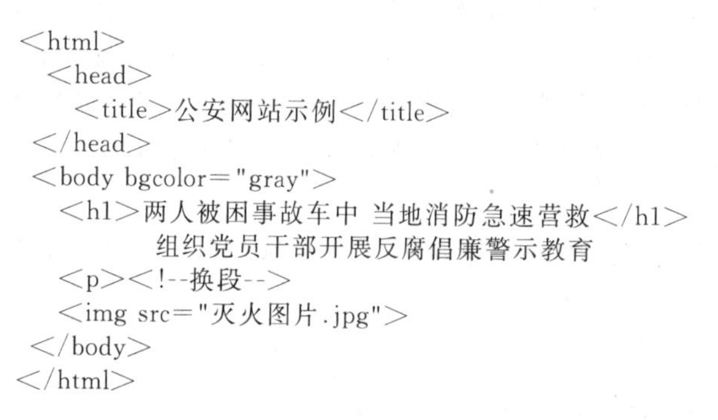

```
<html>
  <head>
    <title>公安网站示例</title>
 </head>
 <body bgcolor="gray">
   <h1>两人被困事故车中 当地消防急速营救</h1>
          组织党员干部开展反腐倡廉警示教育
   <p><!--换段-->
   <img src="灭火图片.jpg">
 </body>
</html>
```

注意事项：HTML 文件对于大小写没有严格限制，可以根据个人习惯自由选择。

代码说明：<html>与</html>标签限定了文档的开始点和结束点，在它们之间的部分是文档的头部和主体。正如上面代码所示，文档的头部由<head>和</head>标签定义，而主体由<body>和</body>标签定义。在文档的头部，可以定义网页标题等信息，如<title>和</title>标签就指明了该网页的标题是“公安网站示例”。在文档的主体部分，可以定义网页的核心内容，如文本、超链接、图像、表格和列表等。本例中只包含了文本和图片，并且用<h1>和</h1>标签定义了“两人被困事故车中　当地消防急速营救”文本为最大标题。在 html 中，<h1> ～ <h6>标签可用于定义标题，<h1>定义最大的标题，<h6>定义最小的标题。html 标记还可以有属性，这些属性位于标记的开始标签中，例如 body 标签表示网页的内容，可以使用 bgcolor 属性设置网页的背景颜色；<img>标签表示插入图片，可以使用 src 属性设置图片的路径和名称。HTML 中也可以有注释，例如<!--换段-->就是一种说明性的注释，注释不显示在最终的网页中，只是用于帮助程序员阅读代码。

2）保存 HTML 文件

将编辑好的 HTML 文件保存到磁盘的目录中，例如保存到 D:\ch01 目录中。文件名“公安网站示例.html”，其中.html 是文件的扩展名。使用记事本编辑 HTML 文件，在保存时，必须将“保存类型”选择为“所有文件”，将“编码”选择为 ANSI。如果保存 HTML 文件时，计算机总是给文件名末尾额外加上.txt，那么在保存 HTML 文件时可以用双引号括起来，如图 1-2 所示。

图 1-2　记事本保存文件名

2. 使用浏览器查看网页

打开浏览器，查看网页。常用的浏览器有 IE、FireFox 等。本书使用 Windows 系统自带的 IE 浏览器(8.0 版本)浏览网页，如果要使用其他的浏览器，请参考相关的文档。打开网页的方法如下：

(1) 打开 HTML 文件所在的文件夹(D:\ch01)，用鼠标双击“公安网站示例.html”文件，系统将启动默认的浏览器，打开这个超文本文件，网页效果与图 1-1 相同。需要注意的是，如果系统安装了其他的浏览器，并且默认的浏览器不是 IE，那么可以使用下面的方法用 IE 浏览器查看网页。

(2) 先打开 IE 浏览器，然后选择“文件”→“打开”命令，在打开对话框中输入文件路径“D:\ch01\公安网站示例.html”，最后单击“打开”对话框中的“确定”按钮，网页效果与图 1-1 相同。

1.1.3　拓展训练

1. HTML 文件以(　　)标签对开始和结束。

A. <body>…</body>　　B. <head>…</head>

C. <html>…</html>　　D. <title>…</title>

2. HTML 标签的属性位于(　　)。

A. 开始标签　　B. 结束标签

C. 开始标签和结束标签之间　　D. A 和 B

3. 关于<title>…</title>标签的说法,错误的是(　　)。

A. 网页正文的开始　　B. 用于显示当前网页的标题

C. 位于<body>…</body>之间　　D. 位于<head>…</head>之间

4. 用来将文本设置为最大标题的 HTML 标签是(　　)。

A. <img>…</img>　　B. <h6>…</h6>

C. <html>…</html>　　D. <h1>…</h1>

5. 制作一个简单的 HTML 网页,标题为"个人网页",要求在浏览器中显示你的姓名、籍贯、出生日期、所在院校、所学专业、所学课程、特长爱好等相关信息,为了增加页面的美观效果可以设置背景颜色、插入图片、建立超链接等。

1.2 CSS

1.2.1 知识要点

CSS(Cascading Style Sheets,层叠样式表)是一系列格式规则,用于控制网页的外观,与 HTML 类似,它也是一种标记语言,不需要编译,直接由浏览器执行。

CSS 在网页制作中被广泛应用,可以控制许多仅使用 HTML 无法控制的属性。例如,可以为所选文本指定不同的字体大小和单位(像素、磅值等);通过使用 CSS 以像素为单位设置字体大小,还可以确保在多个浏览器中以更一致的方式处理网页布局和外观效果。另外,使用 CSS 设置网页格式时,内容与表现形式是相互分开的。网页内容即 HTML 代码位于自身的 HTML 文档中,而定义代码表现形式的 CSS 规则位于 HTML 文档的另一部分(通常为<head>部分)中或另一个文档即外部样式表中。如此一来,网页的外观修改和内容更新就变得相对容易,从而有利于网站的升级和维护。

1.2.2 技能操作

运用 CSS 控制网页外观,具体任务如下:

将如图 1-1 所示网页中的第一行文字即"两人被困事故车中　当地消防急速营救"的显示外观修改为 16 像素、红色、粗体样式,效果如图 1-3 所示。

完成任务的方法有如下两种:

- 通过内部样式表控制网页外观。
- 链入外部样式表控制网页外观。

1. 通过内部样式表控制网页外观

在 HTML 文件的<head>区域里可以直接编写样式表,并借助<style>标签将样式

图 1-3　CSS 控制网页外观

表应用于网页，这种方式可以称为内部样式表，使用该方式完成任务具体步骤如下。

1）修改 HTML 文件的内容

```
<html>
  <head>
    <title>公安网站示例</title>
    <style type="text/css">
    <!--
    h1{
    font-size:16 pixels;
    color:red;
    font-weight:bold;
    }
    -->
    </style>
  </head>
   <body bgcolor="gray">
        <h1>两人被困事故车中　当地消防急速营救</h1>
              组织党员干部开展反腐倡廉警示教育
        <p>
        <img src="灭火图片.jpg">
    </body>
</html>
```

注意事项：有些低版本的浏览器不能识别 style 标记，这意味着低版本的浏览器会忽略 style 标记里的内容，并把 style 标记里的内容以文本直接显示到页面上。为了避免这样的情况发生，可以用加 HTML 注释的方式（<! --注释-->）隐藏内容而不让它显示。

代码说明：CSS 格式规则由两部分组成即选择器和声明。选择器是标识已设置格式元素（如 p、h1、类名称或 id 等）的术语，而声明则用于定义样式元素。上述代码中定义了一个样式 h1（h1 称为选择器），在 h1 中声明了三个属性（用大括号括起来的部分）：font-size 属性描述字体的大小，值为 16 pixels（像素）；color 属性描述字体颜色，值为 red（红色），font-weight 属性描述字体粗细，值为 bold（粗体）。

2）浏览网页效果

打开浏览器，查看网页，具体效果与图 1-3 相同。

2. 链入外部样式表控制网页外观

将样式表保存为一个独立的 CSS 文件，然后在 HTML 文档中用<link>标签链接到这个样式表文件，这种方式可以称为外部样式表，使用该方式完成任务具体步骤如下。

1）编写 CSS 样式表文件

```
h1 {
font-size:16 pixels;
color:red;
font-weight:bold;
}
```

将上面的样式表保存为一个独立的 CSS 文档，可命名为 01.css。

2）修改 HTML 文件的内容

```
<html>
  <head>
    <title>公安网站示例</title>
     <link rel="stylesheet" type="text/css" href="01.css">
  </head>
  <body bgcolor="gray">
    <h1>两人被困事故车中当地消防急速营救</h1>
          组织党员干部开展反腐倡廉警示教育
    <p>
    <img src="灭火图片.jpg">
  </body>
</html>
```

3）浏览网页效果

打开浏览器，查看网页，具体效果与图 1-3 相同。

1.2.3 拓展训练

1. 层叠样式表文件的扩展名为（　　）。

   A. .html　　B. .js　　C. .txt　　D. .css

2. 编写一个 HTML 网页，用来宣传你所就读的学校，并借助 CSS 内部样式表和外部样式表控制外观显示。

1.3 JavaScript

1.3.1 知识要点

HTML 和 CSS 能很好地显示数据，但是不能动态地处理数据，例如进行简单的四则运算、格式验证等。要想在网页里实现动态处理数据，可以借助 JavaScript。

JavaScript 于 1995 年由 Netscape 公司在网景导航者浏览器上首次设计实现而成。因为当时 Netscape 公司与 Sun 公司有合作，Netscape 管理层希望它外观看起来像 Sun 公司的 Java 语言，因此取名为 JavaScript。JavaScript 实际上是一种嵌入在网页中的脚本语言，由浏览器解释执行。使用 JavaScript 能够动态地改变网页的内容，处理用户和浏览器产生的事件。目前，数百万计的网页使用 JavaScript 来实现改进设计、验证表单等功能。

1.3.2　技能操作

运用 JavaScript 对页面数据进行动态处理，具体任务如下：

在网页中对用户输入的手机号码格式进行动态验证：如果用户输入手机号码位数不足 11 位，则单击"提交"按钮后给出提示"手机号必须是 11 位"；如果用户输入手机号码为非数字字符，则单击"提交"按钮后给出提示"手机号只能是数字"，具体效果如图 1-4 和图 1-5 所示。

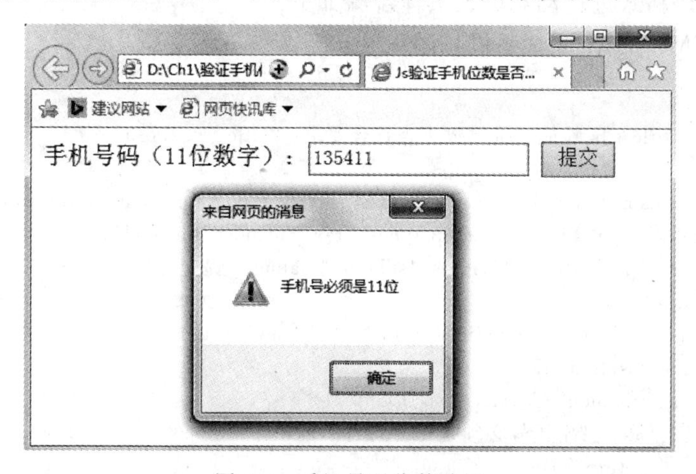

图 1-4　手机号码位数验证

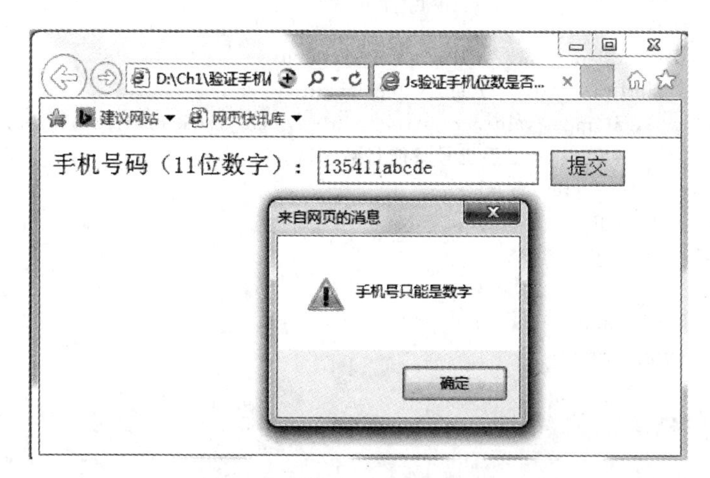

图 1-5　手机号码字符格式验证

完成任务的方法有如下两种：

- 在网页中直接嵌入 JavaScript 代码。
- 引入 JavaScript 文件方式。

1. 直接嵌入方式

通常，在网页中直接嵌入 JavaScript 程序代码的一般格式如下：

```
<script language="text/javascript">
    JavaScript 程序代码
</script>
```

其中的 type 属性表示脚本的类型，其值必须为 text/javascript。JavaScript 程序代码位于<script>和</script>两个标签之间，这些代码由浏览器解释执行。使用这种方式对用户输入的手机号码格式进行动态验证具体步骤如下。

1）编写 HTML 文件“验证手机号码 1. html”

```
<html>
    <head><title>Js 验证手机位数是否符合要求</title></head>
    <body>
        <form name=form1 target="_blank" method=post onsubmit="return dosubmit(this)">
        手机号码(11 位数字): <input type="text" name="mem_id">
        <input type="submit" name="submit1" value="提交">
        </form>
        <script language=" text/javascript ">
        function dosubmit(frm)
        {       if(frm.mem_id.value.length != 11)
                {   alert("手机号必须是 11 位");
                    return false;
                }
                else
                {   var mem_value = frm.mem_id.value;
                    for(var i=0; i<mem_value.length; i++)
                    {   if(mem_value.charAt(i)<'0' || mem_value.charAt(i)>'9')
                        {   alert("手机号只能是数字");
                            return false;
                        }
                    }
                }
                frm.submit();
                return true;
                }
        </script>
    </body>
</html>
```

代码说明：位于<script language=" text/javascript ">和</script>两个标签之间的部分为 JavaScript 程序代码。在这部分代码内部定义了一个名为 dosubmit 的函数，其核心功能是判断用户在文本框中输入的手机号码是否为 11 位数字字符，如果不符合要求，就会弹出相应的提示对话框。

2）浏览网页效果

打开浏览器，查看网页，效果与图 1-4 和图 1-5 相同。

2. JavaScript 文件方式

JavaScript 文件方式是把 JavaScript 程序代码写在一个单独的文件中，然后通过 script 标记把这个 JavaScript 文件引入到网页中。在网页中引入的 JavaScript 文件的具体格式如下：

```
<script type="text/javascript" src="JavaScript 文件的 URL">
</script>
```

其中 src 属性表示要引入的 JavaScript 文件，其值为 JavaScript 文件的 URL。这个 URL 既可以是相对地址，也可以是绝对地址。JavaScript 程序代码位于 JavaScript 文件中，这些代码由浏览器解释执行。需要注意的是，如果 script 标记使用 src 属性引入了一个外部的 JavaScript 文件，那么这个标记的起始标签和结束标签之间的 JavaScript 代码将被忽略，不会被执行。使用这种方式对用户输入的手机号码格式进行动态验证的具体步骤如下。

1）编写 JavaScript 文件 01.js

```
function dosubmit(frm)
  {    if(frm.mem_id.value.length != 11)
       {   alert("手机号必须是 11 位");
           return false;
       }
       else
       {   var mem_value = frm.mem_id.value;
           for(var i=0; i<mem_value.length; i++)
           {   if(mem_value.charAt(i)<'0' || mem_value.charAt(i)>'9')
               {   alert("手机号只能是数字");
                   return false;
               }
           }
       }
       frm.submit();
       return true;
  }
```

2）编写 HTML 文件“验证手机号码 2.html”

```
<html>
```

```
    <head><title>Js验证手机位数是否符合要求</title></head>
    <body>
      <form name="form1" target="_blank" method=post onsubmit="return dosubmit(this)">
      手机号码(11位数字): <input type="text" name="mem_id">
      <input type="submit" name="submit1" value="提交" >
      </form>
      <script language="javascript" src="01.js">
      </script>
    </body>
</html>
```

3）浏览网页效果

打开浏览器,查看“验证手机号码 2.html”网页,效果与图 1-4 和图 1-5 相同。

1.3.3 拓展训练

1. 在 HTML 文件中,JavaScript 程序代码位于(　　)。

   A. <script>和</script>之间　　B. <code>和</code>之间

   C. <javascript>和</javascript>之间　　D. <script>标签内

2. script 标记的 type 属性值为(　　)。

   A. javascript　　B. text　　C. javascript/text　　D. text/javascript

3. 已知 JavaScript 文件 tom.js,文件内容为:

```
alert("tom file");
```

并且在一个 HTML 文档中有下面的代码:

```
<script type="text/javascript" src="tom.js">
    alert("abc");
</script>
```

用 IE 浏览器打开这个 HTML 文档,将弹出一个对话框,对话框中显示(　　)。

   A. tom file abc　　B. tom file　　C. abc　　D. 什么也不显示

4. 将“验证手机号码 2.html”文件中的代码:

```
<script type="text/javascript" src="01.js"></script>
```

改为:

```
<script type="text/javascript" src="01.js">
    alert("我被忽略了,没有执行!!");
</script>
```

保存文件,然后用浏览器打开该文件查看效果,观察“我被忽略了,没有执行!!”这句话是否显示在网页中。

1.4 小　　结

- 网络编程与动态网站制作通常包含客户端开发和服务器端开发两部分：客户端开发主要用于描述信息的内容，也就是浏览器中显示的网页；服务器端开发主要用于实现信息的处理，也就是前后台的动态交互。常用的客户端开发技术有 HTML、CSS、JavaScript 脚本语言等。
- HTML（Hyper Text Markup Language，超文本标记语言）是一种用于描述网页的语言，它不是编程语言，而是标记语言。
- CSS（Cascading Style Sheets，层叠样式表）是一系列格式规则，用于控制网页的外观，与 HTML 类似，它也是一种标记语言，不需要编译，直接由浏览器执行。
- JavaScript 是一种嵌入在网页中的脚本语言，由浏览器解释执行。使用 JavaScript 能够动态地改变网页的内容，处理用户和浏览器产生的事件。

第 2 章

搭建 JSP 动态网站开发环境

JSP(Java Server Pages,Java 服务器端页面)是由 Sun Microsystems 公司(现已被 Oracle 公司收购)倡导多家公司共同参与建立的一种动态网页技术标准。需要说明的是,这里提到的"动态",并不是指放在网页上的动态图片和动画效果,而是指动态交互,即网页会根据用户的要求和选择而动态改变和响应。从本章开始将详细介绍利用 JSP 技术进行网络编程与动态网站制作的具体内容。

2.1 搭建开发平台

2.1.1 知识要点

1. JSP 简介

JSP 技术是在传统的 HTML 网页文件中插入 Java 程序段和 JSP 标记,从而形成 JSP 文件来创建动态页面的一种方法,它使得 Web 服务器、应用程序服务器、浏览器以及各类开发工具环境下的动态网站开发更为便捷。因其具有简单易用、平台无关、安全可靠等特点,目前已成为动态网站制作与开发的主流技术。需要强调的是,在学习 JSP 技术之前,读者应具有较好的 Java 语言基础以及 HTML 语言基础。

2. JSP 引擎

运行包含 JSP 页面的 Web 项目,需要一个支持 JSP 页面运行的 Web 服务软件,该服务软件也被称作 JSP 引擎或 JSP 容器。通常将安装了 JSP 引擎的计算机称为一个支持 JSP 的 Web 服务器。当用户要访问 JSP 页面时,Web 服务器通知 JSP 引擎处理该页面中的元素,返回用户想要的结果,如图 2-1 所示。

目前,比较常用的 JSP 引擎有 Tomcat、JRun、Resin 等,本书采用的是 Tomcat。需要指出的是,JSP 引擎需要 Java 语言的核心库和相应编译器,所以在安装 JSP 引擎之前,需要先安装 J2SE(Java2 Standard Edition,Java2 标准版)提供的 JDK(Java Development Kit,Java 开发工具包),详见 2.1.2 节。

3. Web 服务目录

为了让用户能够通过浏览器访问到 Tomcat 服务器上的 JSP 页面,就必须将写好的 JSP 页面保存到该 Tomcat 服务器的某个 Web 服务目录中,具体包括如下几种情况。

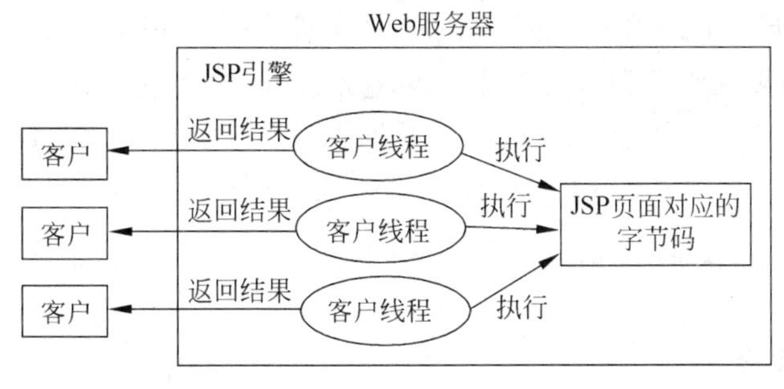

图 2-1　JSP 引擎

1）根目录

如果 Tomcat 的安装目录是 D:\apache-tomcat-6.0.0，那么 Tomcat 服务器的 Web 服务目录的根目录就是 D:\apache-tomcat-6.0.0\webapps\ROOT。

用户如果要访问 Web 服务目录根目录中的 JSP 页面，在浏览器地址栏中依次输入 Tomcat 服务器的 IP 地址、端口号、JSP 页面的文件名即可。例如，Tomcat 服务器的 IP 地址是 192.168.3.1，根目录中存放的 JSP 页面文件名为 first.jsp，那么用户在浏览器地址栏需要输入 http://192.168.3.1:8080/first.jsp。

2）已有的 Web 服务目录

Tomcat 服务器安装目录中的 webapps 目录下的任何一个子目录都可以作为 Web 服务目录。安装 Tomcat 后，webapps 目录下已有 docs、examples、host-manager、manager 子目录，它们都可以作为 Web 服务目录，也可以在 webapps 目录下新建子目录，例如新建 myJSP 子目录，那么 myJSP 就成为一个 Web 服务目录。

如果将 JSP 页面保存到已有的 Web 服务目录中，那么用户访问页面时需要在浏览器地址栏中依次输入 Tomcat 服务器的 IP 地址、端口号、Web 服务目录名、JSP 页面的文件名。例如，Tomcat 服务器的 IP 地址是 192.168.3.1，JSP 页面 first.jsp 保存在 webapps 目录下的 examples 文件夹中，那么用户在访问该页面时就需要在浏览器地址栏中输入 http://192.168.3.1:8080/examples/first.jsp。

3）设置新的 Web 服务目录

可以将计算机中的某个目录设置成一个 Web 服务目录，并为该 Web 服务目录指定虚拟目录，也就是说，隐藏 Web 服务目录的真实位置，而让用户通过虚拟目录访问 Web 服务目录中的 JSP 页面。具体操作是通过修改 Tomcat 服务器安装目录下的 conf 文件夹中的 server.xml 配置文件来完成。例如要将 D:\JSP 和 E:\myPages 设置为新的 Web 服务目录，让用户通过 welcome 和 hello 虚拟目录来访问 D:\JSP 和 E:\myPages 目录中的 JSP 页面，就需要查找 server.xml 配置文件中的</Host>部分（在 server.xml 文件的结尾部分），在其前面加入代码（代码严格区分大小写）：

```
<Context path="/welcome" docBase="D:\JSP" debug="0" reloadable="true" />
<Context path="/hello" docBase="E:\myPages" debug="0" reloadable="true" />
```

配置文件修改后，需要重新启动 Tomcat 服务器才能使设置生效。在这种情况下访问 JSP 页面时，用户需要在浏览器地址栏中依次输入 Tomcat 服务器的 IP 地址、端口号、虚拟目录名、JSP 页面的文件名。例如 Tomcat 服务器的 IP 地址是 192.168.3.1，用户要访问 D:\JSP 或 E:\myPages 中的 JSP 页面 first.jsp，则需要在浏览器地址栏中输入：

http:// 192.168.3.1:8080/welcome/first.jsp 或 http:// 192.168.3.1:8080/hello/first.jsp

2.1.2 技能操作

搭建 JSP 开发平台，具体步骤如下：

- 安装与配置 JDK。
- 安装与配置 Tomcat。
- 编写、保存、运行 JSP 页面。

1. 安装与配置 JDK

1）下载、安装 JDK

（1）登录 http://www.oracle.com/technetwork/java，在 Software Downloads 里选择 Java SE 提供的 JDK，结合本地计算机所使用的操作系统和具体的机器位数，在下载列表中选择相应的 JDK 下载到本地计算机的指定目录。

图 2-2 JDK 安装程序

（2）完成下载后，在目录中会出现如图 2-2 所示的安装程序图标，双击该图标进行 JDK 的安装。

（3）在弹出的许可证协议对话框中单击“接受”按钮，如图 2-3 所示。

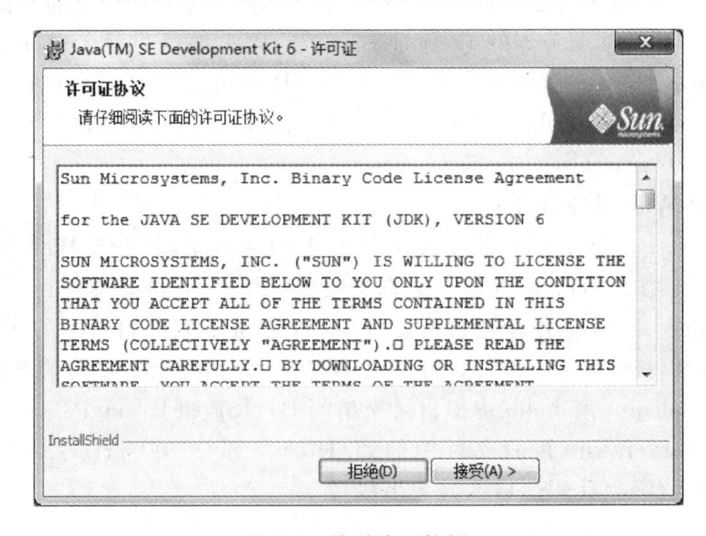

图 2-3 接受许可协议

(4) 在弹出的自定义安装对话框中，默认的安装路径为C:\Program Files\Java\jdk1.6.0\，单击“更改”按钮可以更改安装路径，如图2-4所示。

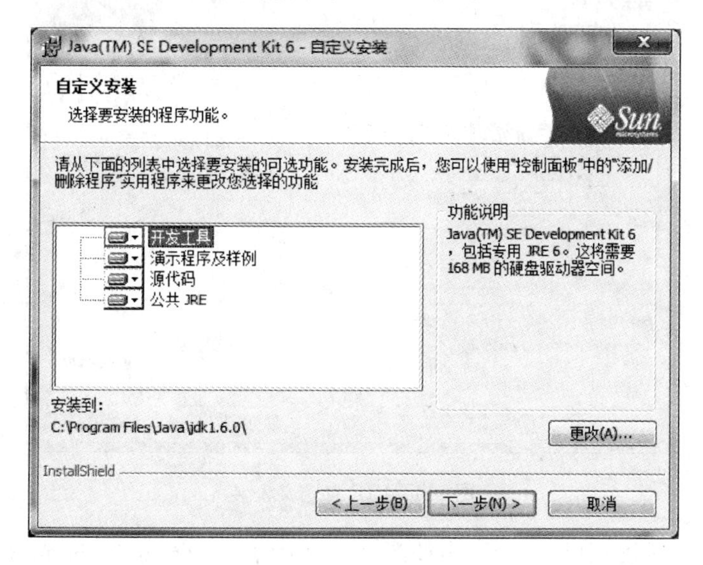

图2-4　自定义安装

(5) 单击“下一步”按钮开始安装，如图2-5所示。

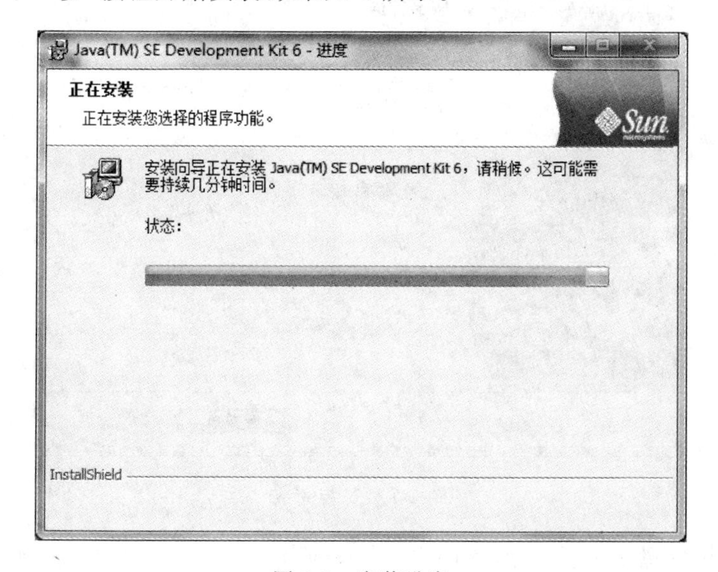

图2-5　安装进度

(6) 进度完成后，弹出自定义安装JRE(Java Runtime Environment，Java运行时环境)对话框，默认的安装路径为C:\Program Files\Java\jre1.6.0\，单击“更改”按钮可以更改安装路径，如图2-6所示。

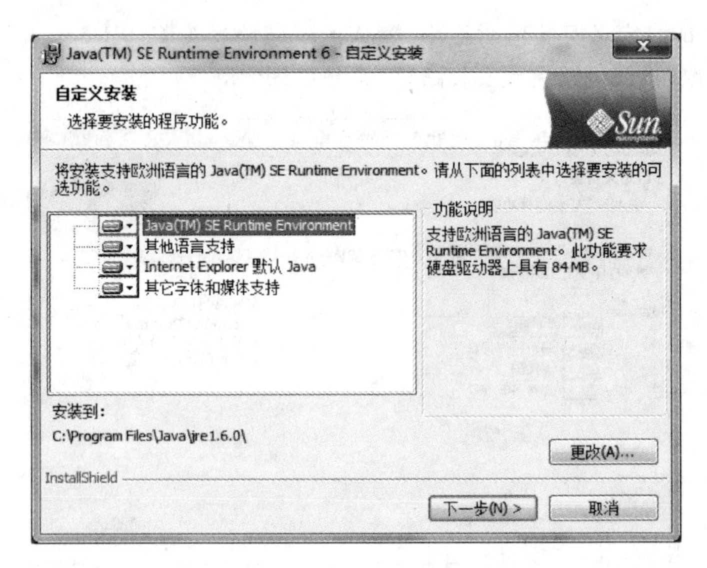

图 2-6 JRE 自定义安装

(7) JRE 安装进度完成后,会弹出如图 2-7 所示的对话框,单击“完成”按钮即可。

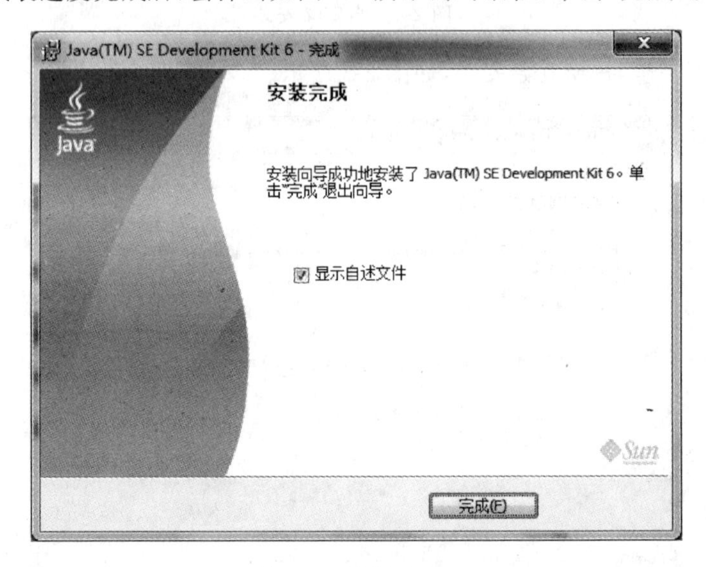

图 2-7 安装完成

2) 配置环境变量 Java_home、Path 和 Classpath

(1) 右击“我的电脑”图标,依次单击“属性”→“高级系统设置”→“高级”→“环境变量”按钮,如图 2-8 所示。

(2) 单击“环境变量”按钮后,弹出如图 2-9 所示的“环境变量”对话框,在这个对话框中可以分别设置环境变量 Path、Classpath、Java_home 的值。

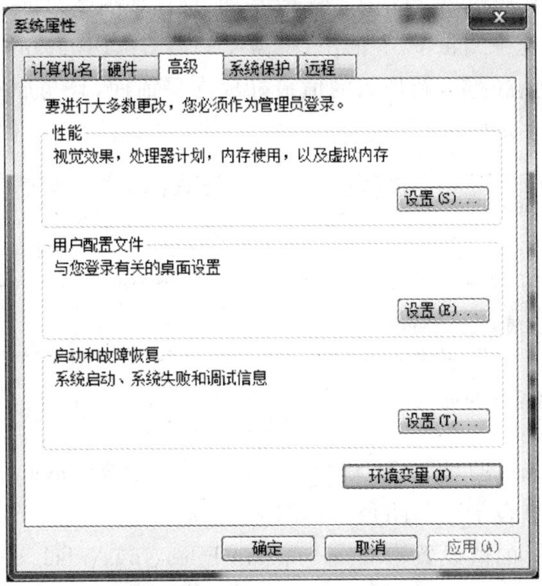

图 2-8　系统属性

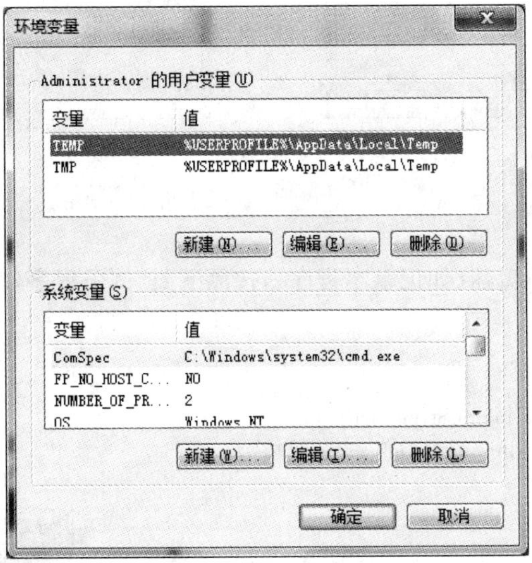

图 2-9　环境变量

在设置前，首先检查“用户变量”和“系统变量”中是否已经有 Path 环境变量存在，如果没有可单击“系统变量”选项组中的“新建”按钮，弹出如图 2-10 所示的“新建系统变量”对话框，在该对话框的“变量名”文本框中输入 Path，在“变量值”文本框中输入 C:\Program Files\Java\jdk1.6.0\bin 即可完成 Path 环境变量的设置；如果事先检查到已经有 Path 环境变量存在，则首先选中 Path，然后单击“编辑”按钮，弹出如图 2-11 所示的“编辑系统变量”对话框。在“变量值”文本框中编辑 Path 的值，将需要的值 C:\Program Files\Java\jdk1.6.0\bin 加入即可，但需要注意的是，新加入的值最好放在最前面，且与原有的值用分号间隔开，以保证 Java 程序和 JSP 页面的正常运行。

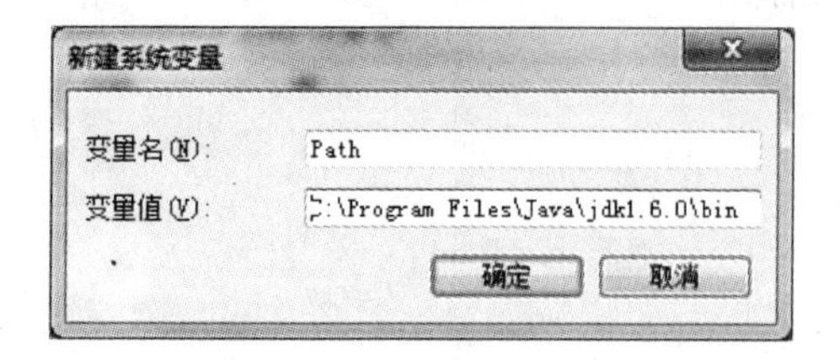

图 2-10　新建系统变量

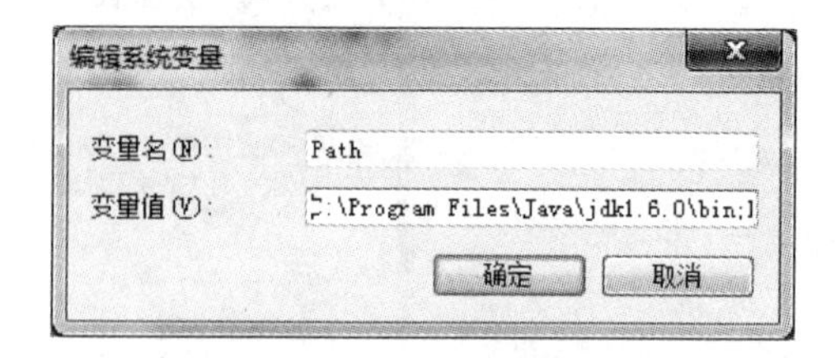

图 2-11　编辑 Path

类似地，将 Classpath 的值设置为“C:\Program Files\Java\jdk1.6.0\jre\lib\rt.jar;.;”即可(其中的“.”表示当前路径)，如图 2-12 所示。

最后，将 Java_home 的值设置为 C:\Program Files\Java\jdk1.6.0，如图 2-13 所示。

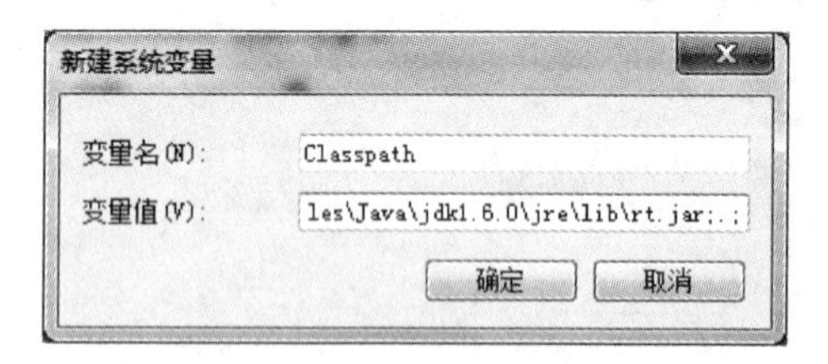

图 2-12　设置 Classpath

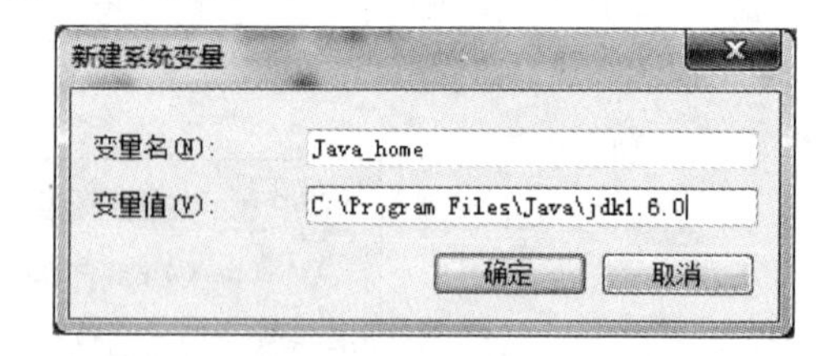

图 2-13　设置 Java_home

3) 测试 JDK

(1) 打开一个文本编辑器(如记事本程序)，按照下面“例子程序代码”的内容输入源程序。

例子程序代码：

```
public class Test{
        public static void main(String args[]){
        System.out.println("测试 JDK,安装成功");
    }
}
```

(2) 将上述源程序保存到 D 盘的某个文件夹中，例如 D:\java。要求保存类型为“所有文件”，文件名为 Test.java。

(3) 在计算机中选择“开始”→“所有程序”→“附件”→“命令提示符”命令，运用 javac 命令编译源文件，运用 java 命令运行程序，如图 2-14 所示。

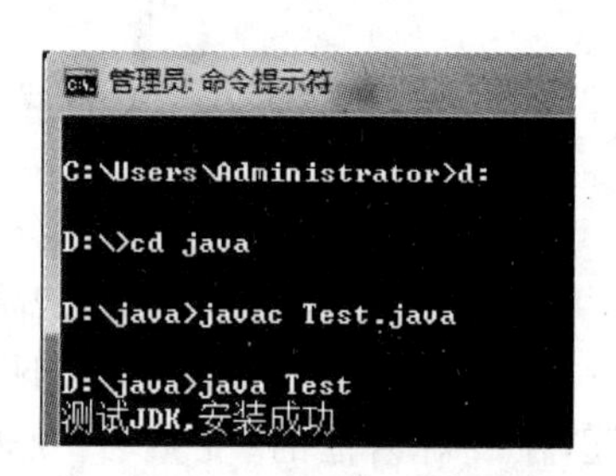

图 2-14　测试 JDK

2. 安装与配置 Tomcat

1）安装 Tomcat 服务器

安装 Tomcat 之前应确保事先已成功安装 JDK，然后登录 http://tomcat.apache.org/下载 Tomcat 程序，然后将其解压到任意盘符下，这里将其解压到 D:\，解压后将出现如图 2-15 所示的目录树结构。

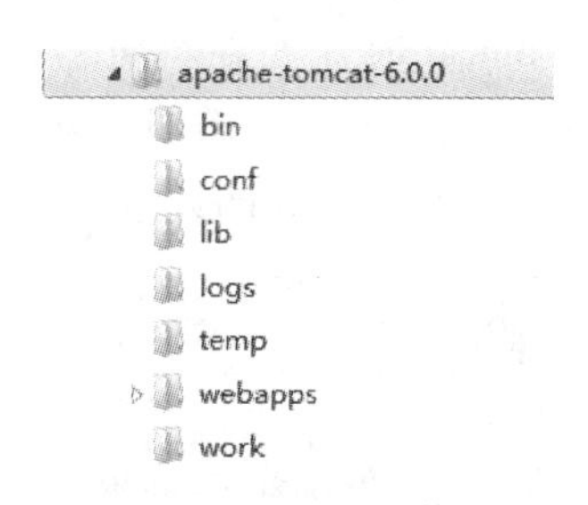

图 2-15　Tomcat 服务器目录树结构

2）启动 Tomcat 服务器

进入 Tomcat 服务器的安装目录，找到 bin 目录中的 startup.bat 批处理文件，然后双击，即可启动服务器。执行 startup.bat 启动 Tomcat 服务器后，会占用一个 MS-DOS 窗口，如图 2-16 所示，如果关闭当前 MS-DOS 窗口也就意味着将关闭 Tomcat 服务器。

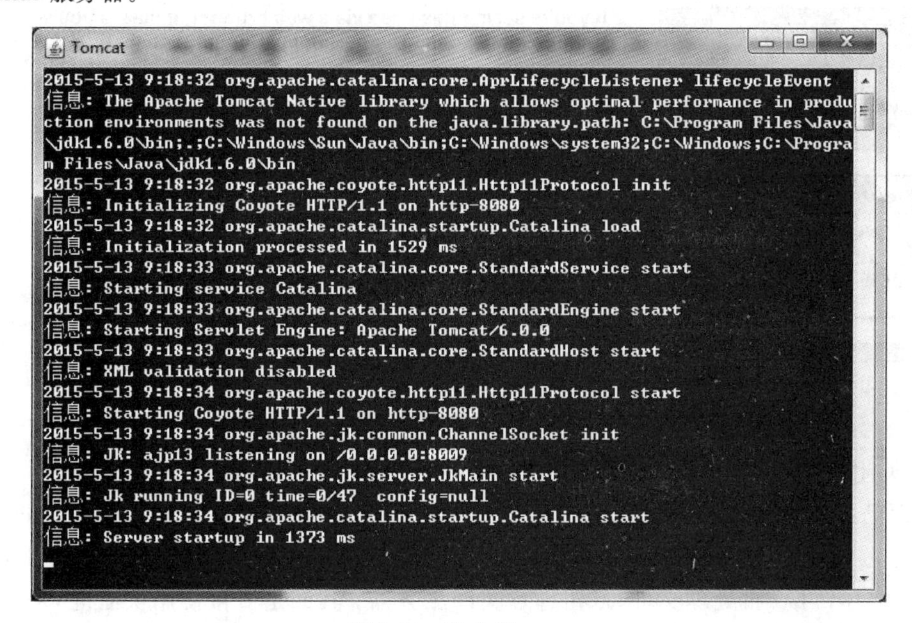

图 2-16　启动 Tomcat

Tomcat 启动后，在浏览器的地址栏中输入 http://localhost:8080 或 http://127.0.0.1:8080，会出现如图 2-17 所示的测试页面。需要说明的是，Tomcat 服务器占用的默认端口号是 8080，如果 Tomcat 所使用的端口已经被占用将会造成 Tomcat 无法正常启动。当然，也可以通过修改 Tomcat 服务器 conf 文件下的主配置文件 sever.xml 来指定 Tomcat 所占用的端口号。用记事本打开 server.xml 文件，查找如下代码：

```
<Connector port="8080" protocol="HTTP/1.1"
           maxThreads="150" connectionTimeout="20000"
           redirectPort="8443" />
```

将其中的 port=8080 更改为新的端口号并重新启动 Tomcat 服务器，例如可将端口号改为 8888 等。Tomcat 安装时默认的端口设置的是 8080，而 http 协议的默认端口是 80，所以测试 Tomcat 时需要在浏览器的地址栏输入 http://localhost:8080 或 http://127.0.0.1:8080，若把 Tomcat 的端口设置为 80，则直接输入 http://localhost 或 http://127.0.0.1 就能显示 Tomcat 测试页面。

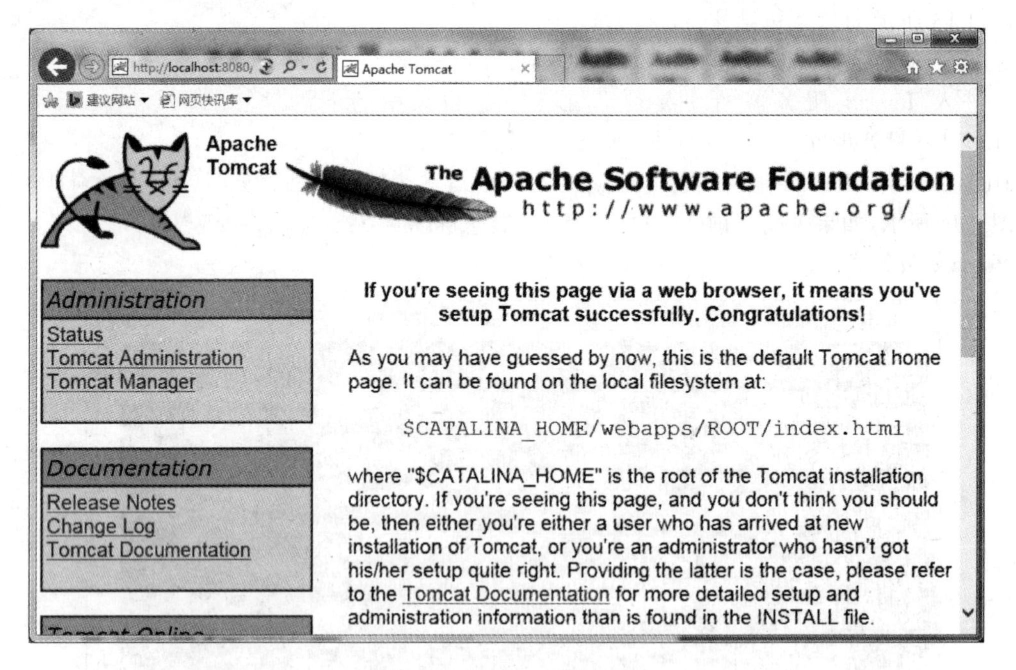

图 2-17　Tomcat 测试页面

3）关闭 Tomcat 服务器

前面已经提到，直接关闭 Tomcat 所占用的 MS-DOS 窗口就意味着关闭 Tomcat 服务器。除此之外，还有两种关闭方法：其一是进入 Tomcat 服务器的安装目录，找到 bin 目录中的 shutdown.bat 批处理文件，双击即可关闭服务器；其二是直接使用快捷键 Ctrl+C 关闭 Tomcat 服务器。

3. 编写、保存、运行 JSP 页面

1）编写 JSP 代码

打开一个文本编辑器。如果是 Windows 操作系统，要打开“记事本”编辑器，可以通过“开始”→“程序”→“附件”→“记事本”命令来完成；如果是其他操作系统，请参考操作系统的帮助手册打开一个文本编辑器。在打开的文本编辑器中输入下面的代码：

```
<%@ page contentType="text/html;charset=GB2312" %>
  <html>
  <head>
```

```
    <title>JSP 页面</title>
  </head>
  <body bgcolor="gray">
  <%
   out.print("这是我的第一个 JSP 页面");
   %>
    <h1>计算 1 到 100 之和</h1>
    <%int i,sum=0;
      for(i=1;i<=100;i++)
         sum+=i;
    %>
    结果: <%=sum%>
  </body>
</html>
```

2）保存 JSP 页面

将编辑的 JSP 代码保存到 Tomcat 服务器的 Web 服务目录，此处选择保存到 Web 服务目录的根目录中，即保存到 D:\apache-tomcat-6.0.0\webapps\ROOT。文件名为 first.jsp，其中.jsp 是文件的扩展名。使用记事本编辑 JSP 文件，在保存时，必须将“保存类型”选择为“所有文件”，将“编码”选择为 ANSI，如图 2-18 所示。

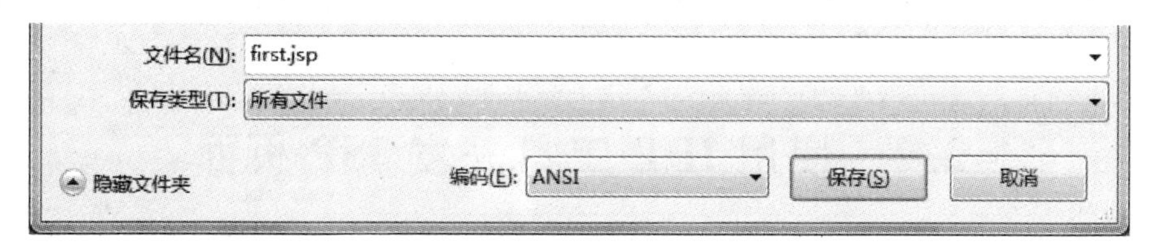

图 2-18　保存 JSP 页面

3）运行 JSP 页面

因为在保存 JSP 页面时，直接保存到了 Web 服务目录的根目录中，即保存到 D:\apache-tomcat-6.0.0\webapps\ROOT，所以在运行 JSP 页面时直接在浏览器地址栏中输入 http://localhost:8080/first.jsp 即可查看页面效果，如图 2-19 所示。

图 2-19　运行 JSP 页面

2.1.3 拓展训练

1. 编写一个 Java 应用程序,要求能够输出 100 以内的奇数。

2. 修改 Tomcat 服务器的主配置文件 server. xml,将默认端口号 8080 修改为 9090,然后测试 Tomcat 服务器是否能够正常启动,打开浏览器在地址栏中输入相应的地址,查看能否出现测试页面。成功后,将端口号改回 8080,重新测试 Tomcat 服务器是否能够正常启动,打开浏览器在地址栏中输入相应的地址,再次查看能否出现测试页面。

3. 如果 E:\myJSP 是一个 Web 服务目录,其虚拟目录为 game,index. jsp 保存在 E:\myJSP\ch02 中,那么在 Tomcat 服务器(端口号为 8080)所在计算机的浏览器地址栏中输入下列哪个地址能够正确访问到 index. jsp 页面?(　　)

A. http://localhost:8080/index. jsp

B. http://localhost:8080/myJSP/ch02/index. jsp

C. http://localhost:8080/game/index. jsp

D. http://localhost:8080/game/ch02/index. jsp

4. 请在 D:\盘建立一个名为 study 的文件夹,并将该目录设置为一个 Web 服务目录,然后用文编辑器编写一个简单的 JSP 页面,保存到该目录中,让用户通过虚拟目录 find 来访问这个 JSP 页面。

2.2 使用 IDE 工具开发 JSP 页面

2.2.1 知识要点

1. JSP 页面

JSP 页面中可以有普通的 HTML 标记,也可以有规定的 JSP 标记,以及通过标记符“<% %>”加入的 Java 程序片段。在保存 JSP 页面时,文件名必须符合标识符的规则,即由字母、数字、下划线和美元符号组成,且不能以数字字符作为开头。一个 JSP 页面文件的扩展名为. jsp。

2. IDE 工具

2.1 节介绍了使用文本编辑器即记事本程序编写 JSP 文件的方法,除了文本编辑器之外,还可以使用 IDE(Integrated Development Environment,集成开发环境)工具来帮助程序员快速开发 JSP 程序,常用的 IDE 工具有 Eclipse、NetBeans、JBuilder 等,本书采用 Eclipse。Eclipse 是由 IBM 等多家公司和组织成立的 Eclipse 基金会开发的开放源代码工具,它本身是一个框架和一组服务,通过插件机制来灵活地构建开发环境。

2.2.2 技能操作

使用 Eclipse 集成开发环境创建 Web 项目,在项目中创建 JSP 文件,然后部署项目到

Tomcat 服务器并运行。具体步骤包括：

(1) 安装启动 Eclipse。

(2) 配置服务器。

(3) 创建 Web 项目和 JSP 文件。

(4) 编写、调试、运行 JSP 页面。

1. 安装启动 Eclipse

登录官方网站 http://www.eclipse.org 下载相应版本的 Eclipse SDK，通常为压缩包格式，解压后双击 eclipse 应用程序图标便可启动进行使用。如果下载的版本为安装版，则需要遵照安装向导指引进行安装后方可启动进行使用。

启动时会弹出如图 2-20 所示的对话框，用于选择工作区，即选择 Web 项目存放的目录位置。

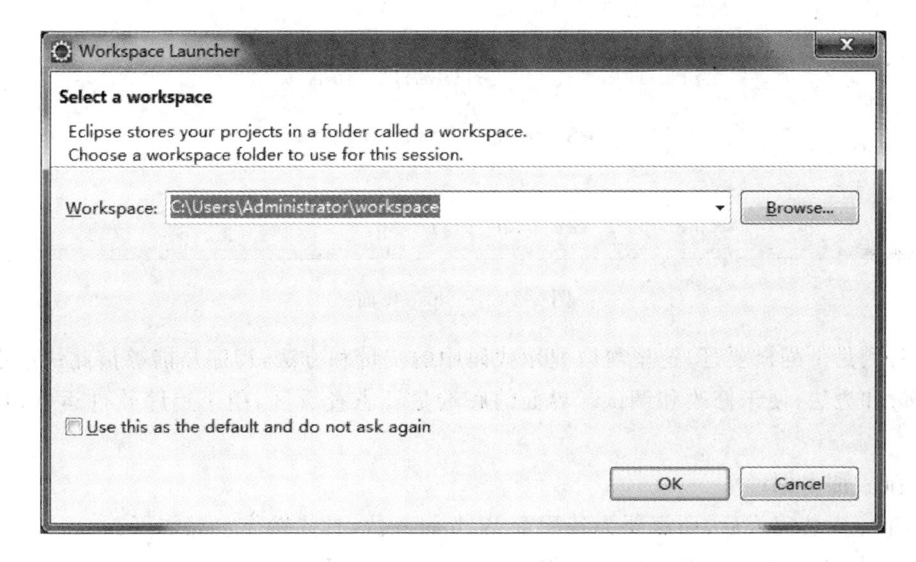

图 2-20　选择工作区

图中默认目录位置为 C:\Users\Administrator\workspace，如果想对此进行更改，可单击 Browse 按钮进行设置。选择工作区之后，单击右下角的 OK 按钮，便可成功启动 Eclipse，启动后的界面如图 2-21 所示。

从图中可以看出，Eclipse 界面的顶端与很多应用软件类似，依次是标题栏、菜单栏、工具栏，标题栏显示的是软件的标题信息，菜单栏中的菜单和工具栏中的工具能够帮助开发人员快速完成所需要的操作。界面的左侧是工程浏览器，展示了在此建立的 Java 项目或 Java Web 项目的树型结构目录，图 2-21 中只有一个 Java Web 项目名为 HelloWorld。界面中间较大的区域是编辑区，在编辑区开发人员可以进行代码的编写，一般默认情况下 Eclipse 会有一些自带的生成代码，实际开发时可以根据需要选择是否留用或全部删除后重新编写。

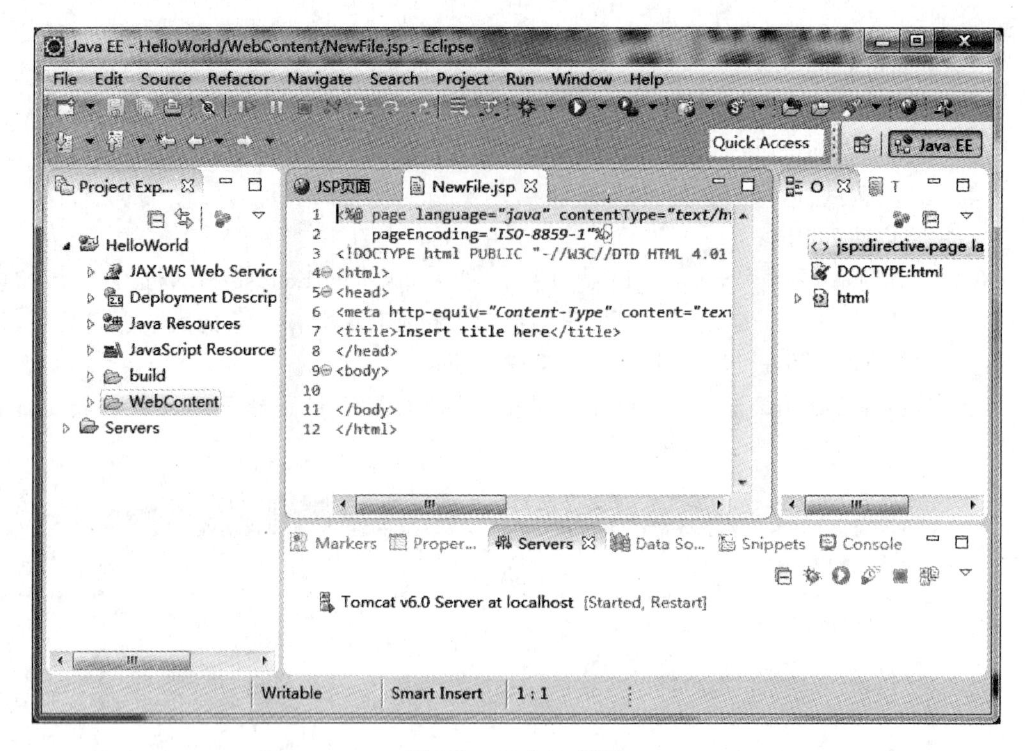

图 2-21 Eclipse 界面

界面的右侧是工程概要区，这里可以显示代码中的变量和方法，程序员能够借此快速定位代码中变量和方法，便于修改和调试。界面的底端是日志查看区，用于追踪软件运行，查看错误日志。

2. 配置服务器

Eclipse 安装完成后，还需要为其配置 Web 服务器，具体操作步骤如下：

(1) 在 Eclipse 窗口中逐级选择 Window → Preferences → Server → Runtime Environment 选项，如图 2-22 所示。

(2) 因为此前从未进行过服务器的相关设置，所以需要单击右侧的 Add 按钮，来添加一个新的服务器运行时环境。前面介绍过，我们已经安装了 Tomcat 6.0 作为 Web 服务器，所以在此选择 Apache Tomcat v6.0，如图 2-23 所示。

(3) 单击 Next 按钮，在弹出的对话框中单击 Browse 按钮，可以在接下来弹出的“浏览文件夹”对话框中选择 Tomcat 的安装目录，在此选择的是 D:\apache-tomcat-6.0.0，如图 2-24 所示。

(4) Tomcat 安装目录选好后单击“确定”按钮即可。在该对话框中继续单击 Install JREs(安装 Java 运行时环境等)按钮，会弹出如图 2-25 所示的对话框，在该对话框中进行后续 JRE 的设置。

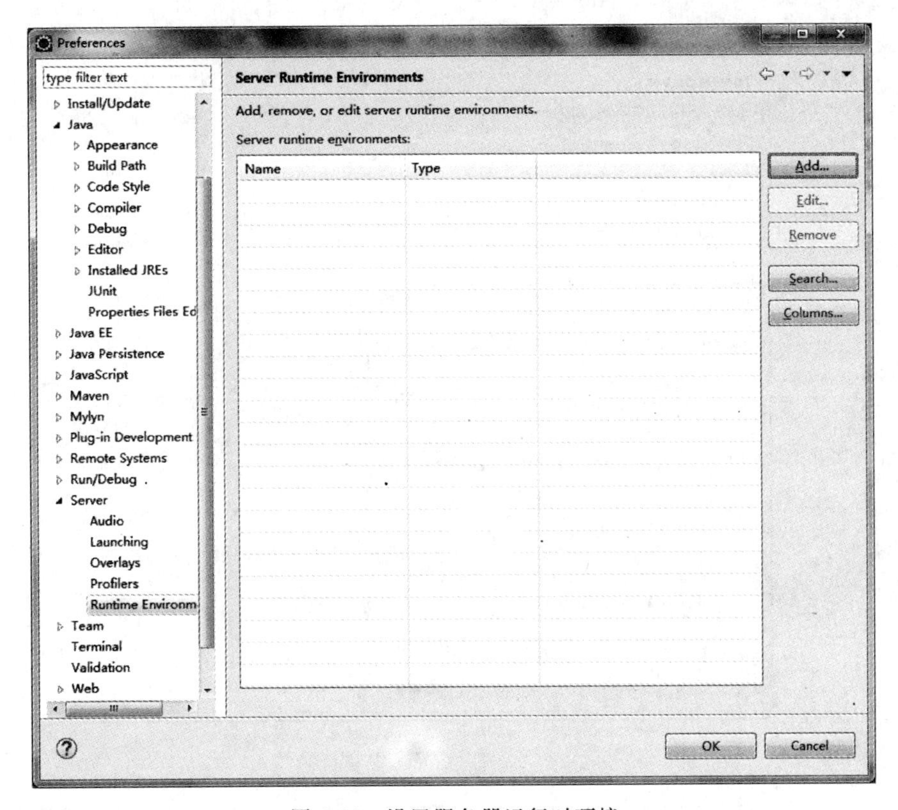

图 2-22　设置服务器运行时环境

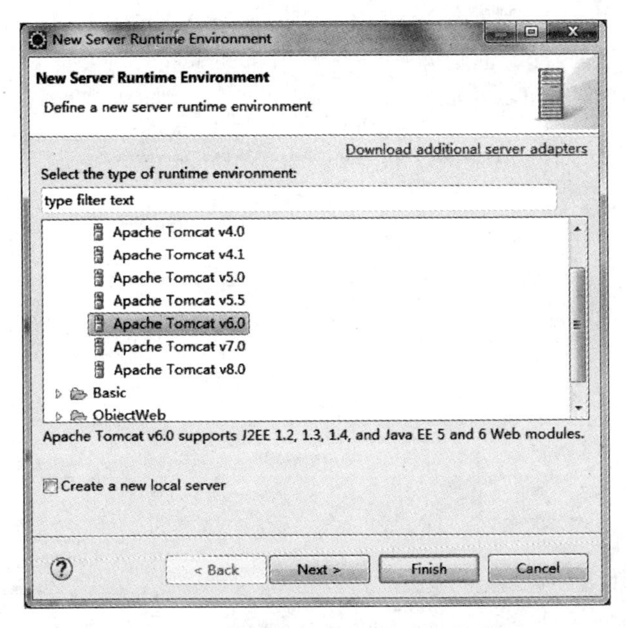

图 2-23　添加新的服务器运行时环境

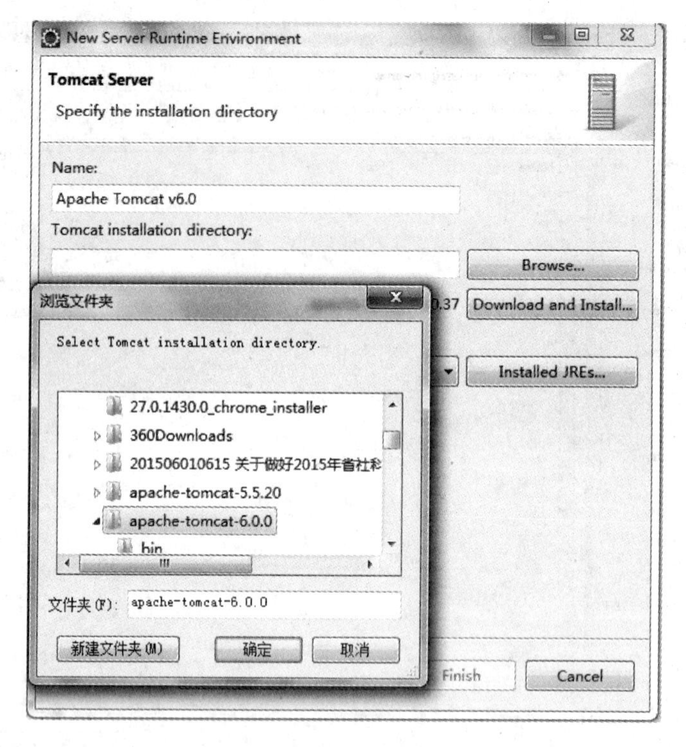

图 2-24　选择 Tomcat 安装目录

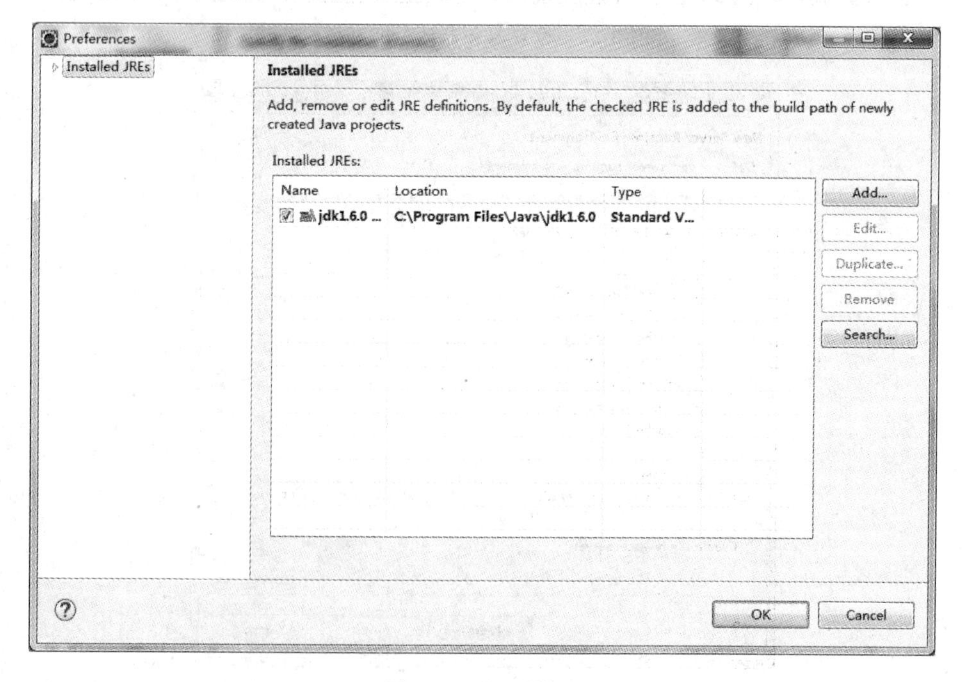

图 2-25　JRE 设置

(5) 如果这里已经配置了JDK,可以在此直接进行选择,同时也可以单击右侧的Edit、Duplicate、Remove、Search按钮分别进行编辑、复制、移除、搜索等相关操作。如果之前没有配置过JDK,则需单击右侧的Add按钮,新增一个标准的JRE,如图2-26所示。

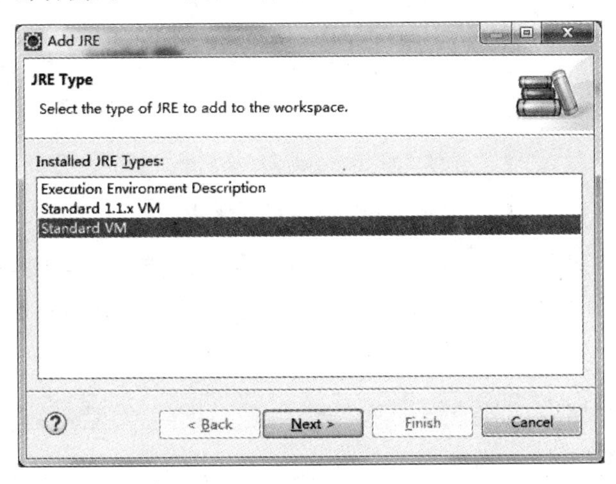

图2-26　新增标准的JRE

(6) 在如图2-26所示的添加JRE对话框中单击Next按钮后,弹出如图2-27所示的JRE安装目录选择界面,单击该界面中JRE home文本框后面对应的Directory按钮,选择好目录然后单击"确定"按钮,在此选择的是C:\Program Files\Java\jdk1.6.0。

经过上述步骤,Web服务器的配置已经完成。

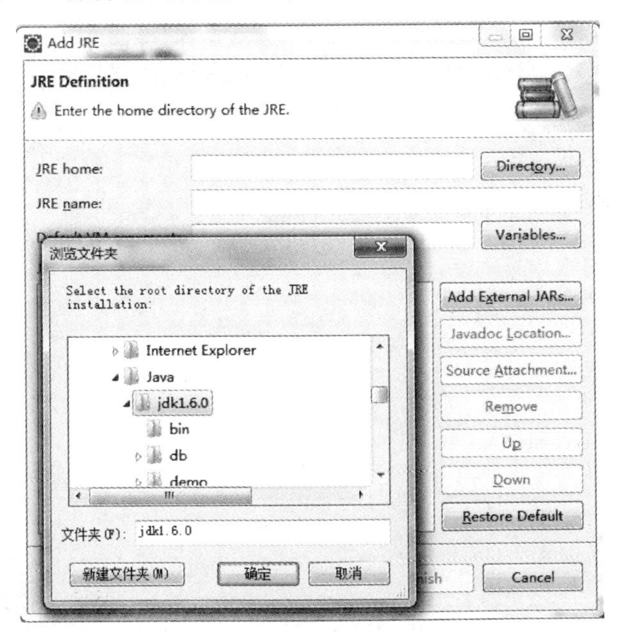

图2-27　选择JRE安装目录

3. 创建 Web 项目和 JSP 文件

(1) 进行完服务器的配置，接下来就可以创建 Java Web 项目了。在 Eclipse 主窗口逐级选择 File→New→Dynamic Web Project 菜单命令来创建项目，如图 2-28 所示。

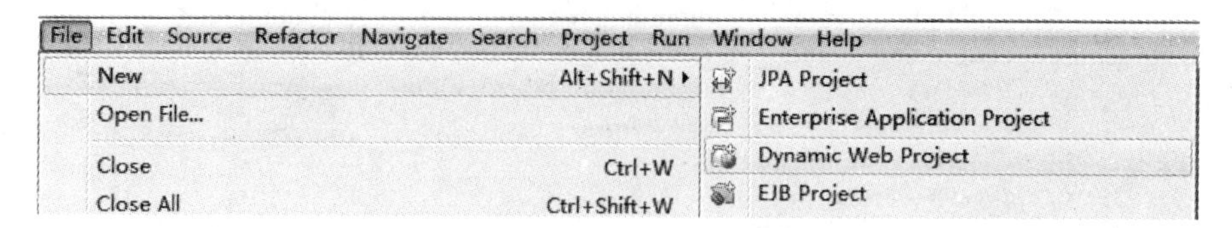

图 2-28 创建 Web 项目

(2) 在弹出的对话框中，可以进行 Web 项目的详细设置，如在 Project name 文本框中为项目进行命名，本例中命名为 HelloWorld；同时在 Target runtime 下拉列表中选择之前配置过的服务器 Apache Tomcat v6.0，如图 2-29 所示。

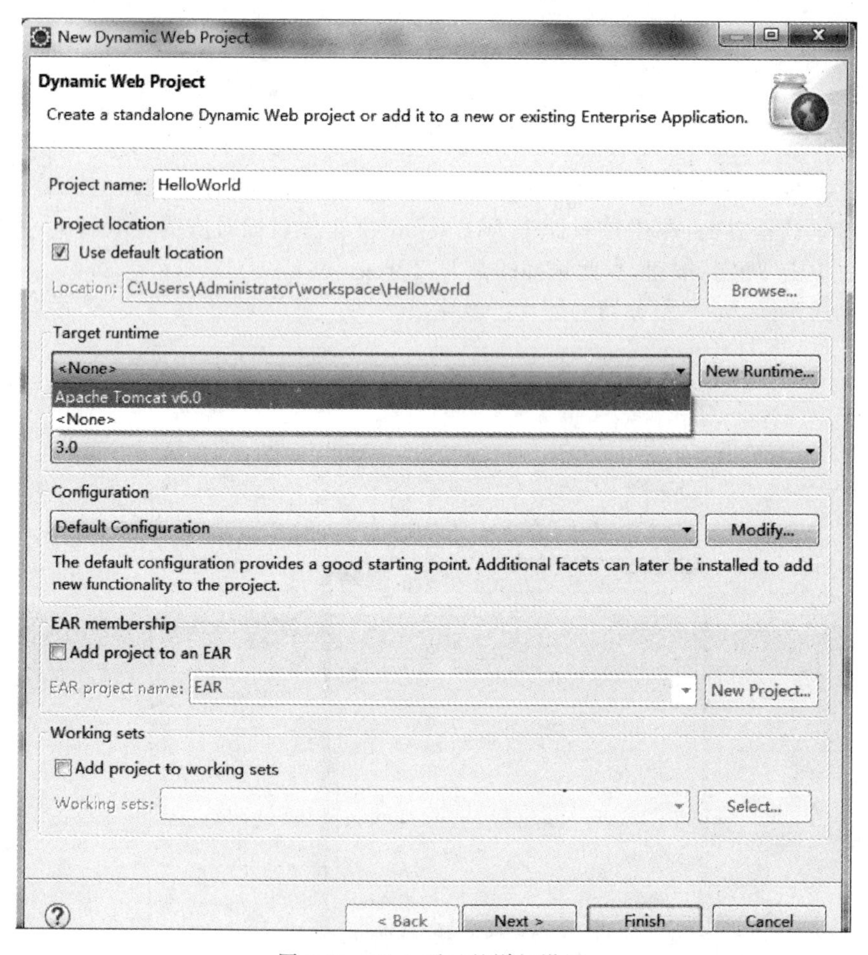

图 2-29 Web 项目的详细设置

(3) 单击 Next 按钮后,进入下一步,按照向导提示进行完所有的设置后单击 Finish 按钮完成 Web 项目的创建。项目创建后,在 Eclipse 窗口左侧目录树中就会出现对应的目录结构,在该结构中的 WebContent 节点上右击,依次选择 New→JSP File 命令,新建一个 JSP 文件,如图 2-30 所示。新建的 JSP 文件,默认名称为 NewFile. jsp,可以根据实际需求对该文件名进行修改。

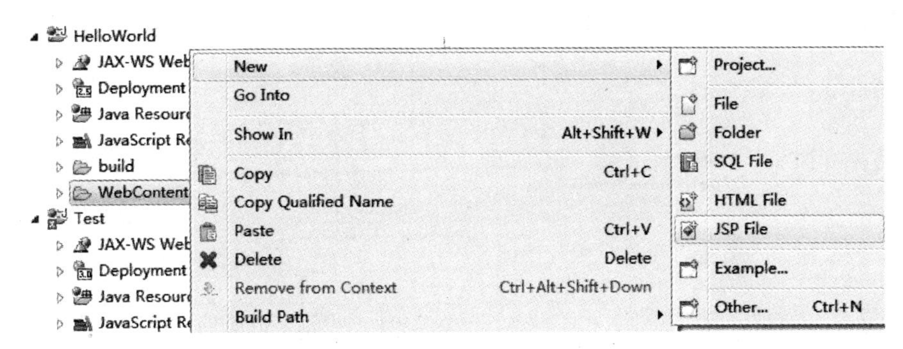

图 2-30　创建 JSP 文件

4. 编写、调试、运行 JSP 页面

1) 编写代码

在新建的 JSP 文件中,会自动生成部分代码,开发人员只需要在核心部分编写关键代码即可。下面是一个简单的 JSP 文件,文件名为 NewFile. jsp,其中只有<body>与</body>标记间的内容为开发人员手动输入,其余代码均为自动生成。

```
<%@ page language="java" contentType="text/html; charset=ISO-8859-1"
    pageEncoding="ISO-8859-1"%>
<!DOCTYPE html PUBLIC "-//W3C//DTD HTML 4.01 Transitional//EN"
"http://www.w3.org/TR/html4/loose.dtd">
<html>
<head>
<meta http-equiv="Content-Type" content="text/html; charset=ISO-8859-1">
    <title>Insert title here</title>
</head>
<body>
HelloWorld!
This is my first JSP file!
</body>
</html>
```

2) 调试、运行 JSP 页面

编写好的代码如果有语法错误,需要手动修改。如果没有错误,则在 Eclipse 左侧目录结构中找到该文件,然后右击,依次选择 Run As→Run on Server 菜单命令,如图 2-31 所示。在打开的 Run On Server 对话框中,选中 Always use this server when running this project 复选框，其他选用默认设置，如图 2-32 所示。

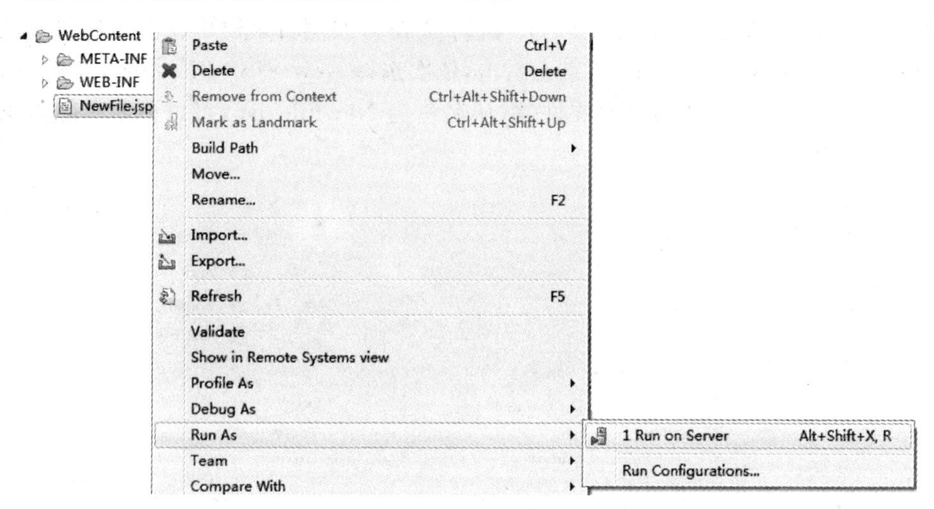

图 2-31　运行 JSP 页面的方法

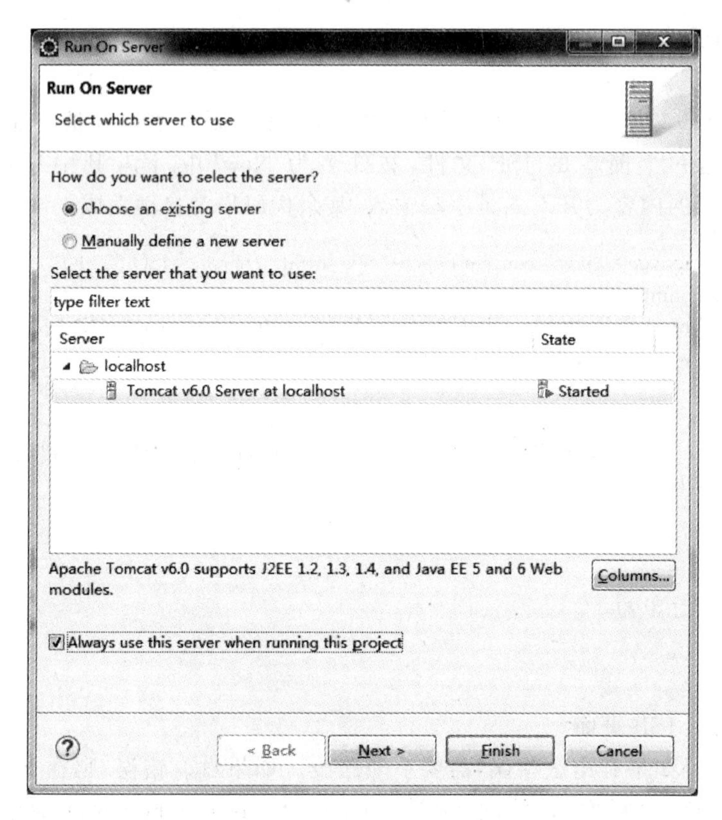

图 2-32　Run on Server 对话框

前面提到的 NewFile.jsp 文件代码，在 Eclipse 运行后的效果如图 2-33 所示。

图 2-33　查看 JSP 页面运行效果

需要注意的是，如果在浏览器中运行上面的 JSP 页面，则需要在浏览器地址栏中输入 http://localhost:8080/HelloWorld/NewFile.jsp，也就是说，Eclipse 建立的项目 HelloWorld 会被认为是一个 Web 服务目录。

2.2.3　拓展训练

1. 参考本节技能操作中创建 Web 项目和 JSP 文件的步骤，创建一个名为 Test 的 Java Web 项目，并在该项目中创建一个名为 myTest.jsp 的文件。
2. 参考本节技能操作中编写、调试、运行 JSP 页面的步骤，在 myTest.jsp 页面中显示“这是我创建的 JSP 文件，全都学会了，好高兴啊！”，并将其发布运行。
3. 思考如何将已有代码导入到 Eclipse 中，并试着完成操作。

2.3　小　　结

- JSP 技术不仅是开发 Java Web 的主流技术，而且也是进一步学习相关技术（如模式和框架）的基础。
- JSP 引擎是支持 JSP 程序的 Web 服务器，负责运行 JSP 程序，并将相关结果发送给客户端。目前流行的 JSP 引擎有 Tomcat、Resin、JRun、WebSphere、WebLogic 等，本书使用的是 Tomcat 服务器。
- 当服务器上的一个 JSP 页面第一次被客户请求执行时，服务器上的 JSP 引擎首先将 JSP 文件转译成一个 Java 文件，并将 Java 文件编译成字节码文件，然后执行字节码文件响应客户端的请求。
- 安装 Tomcat 服务器，首先安装 JDK，并需要设置 Java_home 环境变量。

第 3 章

JSP 基本语法

本章将 JSP 所包含的各类元素，如声明、注释、程序片、表达式、JSP 指令标记(Directive Elements)、JSP 动作标记(Action Elements)等主要方面来说明 JSP 的基本语法结构。

3.1 JSP 文件的构成

3.1.1 知识要点

在传统的网页即 HTML 文件中插入 Java 动态元素、JSP 标记等，就形成了 JSP 页面。通常，一个 JSP 文件包括如下组成部分：

(1) HTML 标记，这在第 1 章中已有介绍。

(2) Java 动态元素，包括变量和方法的定义、Java 程序片和 Java 表达式。

(3) 注释，包括 HTML 注释和 JSP 特有的注释。HTML 注释格式为：<! --注释内容-->，JSP 特有的注释格式为：

```
<%--注释内容--%>
```

(4) JSP 标记，包括 JSP 指令标记和 JSP 动作标记。

3.1.2 技能操作

识别 JSP 文件中的构成要素，具体任务如下：

- 阅读下面的代码，识别其中的 JSP 构成要素。
- 分别在记事本和 Eclipse 中输入下面的代码，并对 JSP 页面进行调试运行。

1. 阅读下面的代码，识别其中的 JSP 构成要素。

代码模板 eg3_1.jsp 如下：

```
<%@ page language="java" contentType="text/html; charset=GBK"
pageEncoding="GBK"%><!-- JSP 指令标记 -->
<%--下面是变量的声名和方法的定义--%>
<%!
    int i=0; //变量声明
    double mul(double a,double b){ //方法定义
        return a * b;
```

```
    }
%>
<html>
    <head>
        <title>JSP 文件构成</title>
    </head>
     <body bgcolor="cyan"><!--HTML 标记 -->
    <%i++;
    double mulitiple=mul(5.5,6.0); //Java 程序片
   %>
  第
  <%=i %>
  次计算,结果为
  <%=mulitiple %><!--Java 表达式 -->
 </body>
</html>
```

2. 分别在记事本和 Eclipse 中输入上面 eg3_1.JSP 中的代码,并对 JSP 页面进行调试运行。

1) 使用记事本和浏览器完成

按照 2.1 节介绍的方法,打开记事本编写代码,然后将代码保存到自己设定的 Web 服务目录(如 ch03)中,最后在浏览器地址栏输入相应的地址访问该 JSP 页面,完成效果如图 3-1 所示。

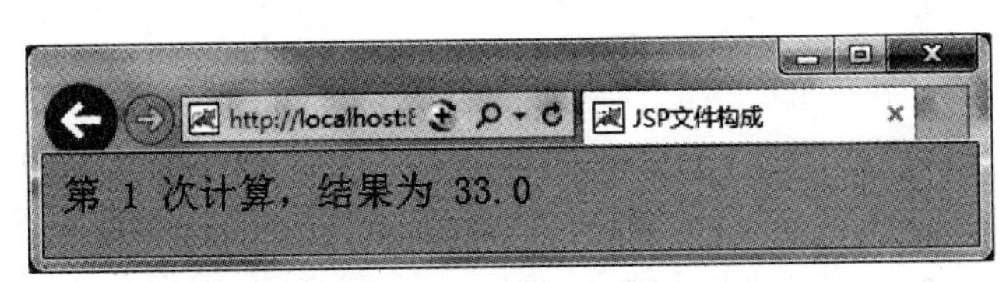

图 3-1 JSP 文件构成页面效果

2) 使用 Eclipse 完成

按照 2.2 节介绍的方法,在 Eclipse 中建立 Web 项目和 JSP 文件,然后编写并保存代码,最后在 Eclipse 中完成运行,效果同图 3-1。

3.1.3 拓展训练

1. 在 JSP 文件中,方法的定义使用(　　)符号。

   A. <% %>　　B. <%! %>　　C. <%@ %>　　D. <%@ %>

2. 在 JSP 文件中,Java 程序片使用(　　)符号。

   A. <% %>　　B. <%! %>　　C. <%@ %>　　D. <%@ %>

3. 在 Tomcat 服务器下新建服务目录 ch03,打开记事本,输入如下的 Test3_1.jsp 程

序，并启动 Tomcat 服务器，在地址栏中输入 http://localhost:8080/ch03/Test3_1.jsp 访问 JSP 页面。

代码 Test3_1.jsp 如下：

```
<%@ page contentType="text/html;charset=gb2312" %>
<html>
	<body bgcolor="cyan">
	<h3>求两个数之和</h3>
	<!--下面是求两个数之和的方法的声明过程-->
	<%!
		double sum(double a,double b){
			return a+b;
		}
	 %>
	<%/*下面是方法的调用过程,并且注释的内容在客户端看不到*/%>
	<%= sum(3.3,4.5)%>
	</body>
</html>
```

3.2 变量和方法的定义

3.2.1 知识要点

1. 变量的定义

JSP 文件中，可以在“<%!”和“%>”符号之间定义成员变量，变量类型可以是 Java 语言所支持的任何数据类型。例如下面的例子中定义了两个普通型变量 i 和 j 初值都为 0，定义了一个双精度实型变量 k 初值为 0.0，定义了一个字符串对象 s 其实体为“hello”。

```
<%!
   int i=0,j=0;
   double k=0.0;
   String s="hello";
%>
```

变量的定义在书写位置上没有固定要求，但习惯上总是将其放在 JSP 文件的前端。成员变量一旦定义就在整个 JSP 页面内都有效，且为多个用户所共享。也就是说，任何用户操作该变量都会对其他用户产生影响，原理如图 3-2 所示。例如 3.1.2 节中的 JSP 文件 eg3_1.jsp，第一个用户访问页面时显示效果为“第 1 次计算，结果为 33.0”，第二个用户再访问页面时显示效果则为“第 2 次计算，结果为 33.0”，以此类推；原因就在于成员变量 i 为多个用户所共享。

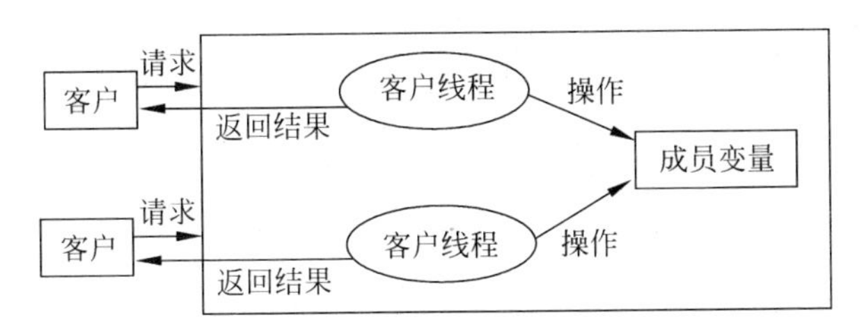

图 3-2　共享成员变量

2. 方法的定义

JSP文件中，也可以在“<%!”和“%>”符号之间定义方法，这些方法将在Java程序片中被调用以完成相应的功能，关于Java程序片的具体内容会在下一节中详细介绍。需要强调的是，这些方法在整个JSP页面内都有效，但方法内部定义的变量为局部变量只在该方法内部有效。例如下面的例子就是定义了一个名称为mul的方法，用于求两个实数的乘积，返回值类型为双精度实型，其中的变量a和b为局部变量，只在该方法内部有效，离开了这个方法，便不能再被访问。

```
<%!
    double mul(double a,double b){
        return a * b;
%>
```

3.2.2　技能操作

编写一个JSP页面文件eg3_2.jsp，要求在该页面文件中定义四个成员变量：result1、result2、result3和result4(初值均为0)；然后定义四个方法，分别为add(求两个数的和)、sub(求两个数的差)、mul(求两个数的积)和div(求两个数的商)。另外，还要求在页面文件中编写一段Java程序片，在Java程序片中调用上述四个方法以实现两个数的加、减、乘、除运算，并将最终结果在页面上显示出来。

代码模板eg3_2.jsp如下：

```
<%@ page language="java" contentType="text/html; charset=GBK"
    pageEncoding="GBK"%>
<html>
<head>
<title>变量和方法的定义</title>
</head>
<%!
    double result1,result2,result3,result4;
    double add(double a, double b){
```

```
        double c=a+b;
        return c;
        }
      double sub(double a, double b){
        double c=a-b;
        return c;
        }
      double mul(double a, double b){
        double c=a*b;
        return c;
        }
      double div(double a, double b){
              double c;
              if(b!=0) {
                 c=a/b;
                   return c;
                   }
              else return 0;
        }
%>
<body bgcolor="cyan">
<%
      result1=add(3.5,1.2);
      result2=sub(3.5,1.2);
      result3=mul(3.5,1.2);
      result4=div(3.5,1.2);
      out.print("两个数的和是："+result1+",");
      out.print("两个数的差是："+result2+",");
      out.print("两个数的积是："+result3+",");
      out.print("两个数的商是："+result4+".");
%>
</body>
</html>
```

运行效果如图 3-3 所示。

http://localhost:8080/Test/eg3_2.jsp

两个数的和是：4.7,两个数的差是：2.3,两个数的积是：4.2,两个数的商是：2.916666666666667。

图 3-3　变量和方法的定义页面效果

3.2.3　拓展训练

1．编写 JSP 页面，完成显示 1＊2＊3＊…＊10 乘积的功能，最终效果如图 3-4 所示。

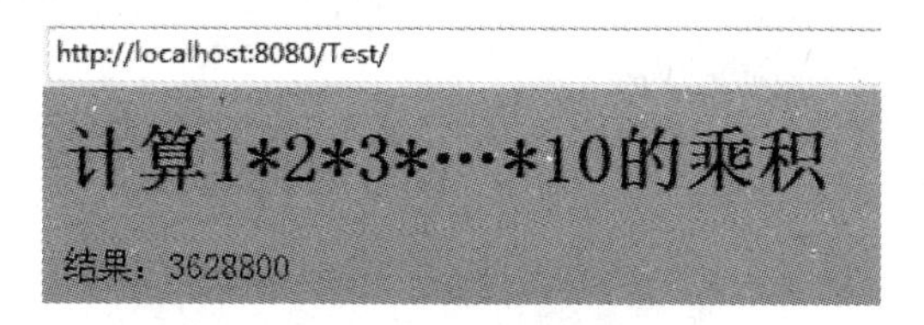

图 3-4　累乘页面效果

要求将如下代码模板补充完整，并进行调试运行。

```
<%@ page contentType="text/html;charset=GB2312" %>
    <html>
    <head>
      <title>练习 1</title>
    </head>
    <body bgcolor="gray">
       【代码段 1】//定义长整型变量 m，初值为 1；定义整型变量 i，初值为 1
      <h1>计算 1 * 2 * 3 * … * 10 的乘积</h1>
      <%
        while(i<=10){
            【代码段 2】//给出累乘的算法
        }
      %>
      结果：<%=m%>
    </body>
</html>
```

2．编写 JSP 页面，完成求和及求平均值的功能，并在页面中显示，最终效果如图 3-5 所示。

http://localhost:8080/Test/

求和结果：15.0 求平均值结果：3.0

图 3-5　求和及求平均值页面效果

要求将如下代码模板补充完整，并进行调试运行。

```
<%@ page contentType="text/html;charset=GB2312" %>
    <html>
    <head>
      <title>练习 2</title>
    </head>
    <body bgcolor="gray">
    <%!
     float sum(float[] array){
        float sum=0.0f;
        for(int i=0;i<array.length;i++)
```

```
          {
              sum+=array[i];
          }
          return sum;
        }
        【代码段 1】//定义求平均值的方法
    %>
    <%
      float[] a={1,2,3,4,5};
      float s=sum(a);
      float result=aver(s,a.length);
     %>
      求和结果:<%=s%>
      求平均值结果:<%=result%>
    </body>
</html>
```

3.3 Java 程序片

3.3.1 知识要点

JSP 文件中,可以在"<%"和"%>"符号之间插入 Java 程序代码,使得 JSP 页面的功能更加完备,这些程序代码在 JSP 页面中被称为 Java 程序片或 Java 程序段。一个 JSP 页面可以有很多个 Java 程序片,这些程序片将被 JSP 引擎(即 Tomcat 服务器)按顺序执行。在程序片中声明的变量为局部变量,具体的作用范围与其声明位置有关,即程序片中的局部变量从声明之处起,在 JSP 页面后续的所有 Java 程序片及 Java 表达式内部都有效,原理如图 3-6 所示。

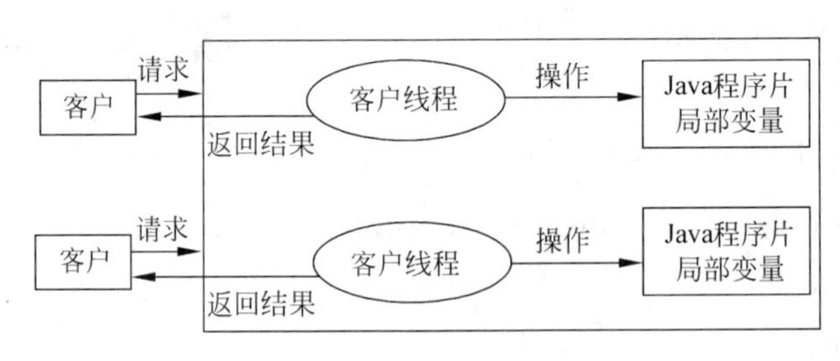

图 3-6 Java 程序片局部变量

3.3.2 技能操作

编写程序产生 1~7 之中的任意一个随机数,判断随机数是几,并相应地输出星期几。具体步骤如下:

- 用JSP声明方式产生一个1～7的随机数方法。
- 用JSP程序片判断产生的随机数是几。
- 判断以后在页面中相应地显示出星期几。

打开记事本，输入下面程序，并保存到相应的Web服务目录。

代码模板 eg3_3.jsp 如下：

```
<%@ page language="java" import="java.util.*" pageEncoding="GB2312"%>
<html>
  <body bgcolor="cyan">
  <h1>
  <!--声明random方面，产生一个1到7的随机数 -->
  <%!
    int random(){
      int i;
      Random r = new Random();
      i=r.nextInt(7)+1;
      return i;
  }
  %>
 <!--下面程序是对random方法的调用，并判断输出星期几 -->
  <%
   int i = random();
   out.print("本次访问产生的随机数是:"+i+"<br>");
   switch (i){
      case 1:
             out.print("星期一");
             break;
      case 2:
             out.print("星期二");
             break;
      case 3:
             out.print("星期三");
             break;
      case 4:
             out.print("星期四");
             break;
      case 5:
             out.print("星期五");
             break;
      case 6:
             out.print("星期六");
             break;
      case 7:
             out.print("星期日");
```

```
            break;
    }
     %>
    </h1>
  </body>
</html>
```

运行效果如图 3-7 所示。

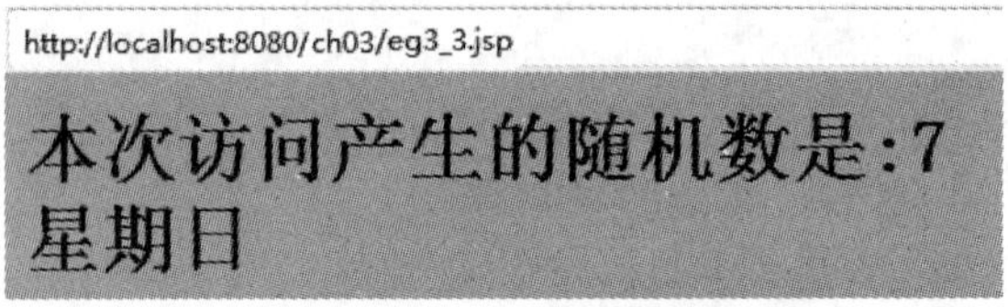

图 3-7　显示星期页面效果

3.3.3　拓展训练

1. 声明一个求梯形面积的方法,给定上底、下底和高求梯形面积,最终运行效果如图 3-8 所示。

图 3-8　求梯形面积页面效果

要求将如下代码模板补充完整,并进行调试运行。

```
<%@ page contentType="text/html;charset=gb2312" %>
<html>
<body bgcolor="gray">
  <h1>
    <%!
    double area;
    double getArea(double a,double b,double c){
        【代码段 1】//求梯形的面积公式
        return area;
    }
     %>
  给定的梯形的上底、下底和高分别是: 3,5,4
  <br>
  面积是:
```

```
  <%=【代码段2】//获得梯形面积
  %>
  </h1>
</body>
</html>
```

2. 模仿上面的例题完成求三角形面积的功能(注意：前提是给定的三条边能够构成三角形)。

3. 模仿上面的例题完成求圆形面积的功能。

3.4　Java表达式

3.4.1　知识要点

JSP文件中,可以在“<%=”和“%>”符号之间插入Java表达式,这个表达式必须拥有确定的值,其值的计算由Web服务器完成,计算结果会以字符串的形式发送到客户端浏览器进行显示。需要注意的是,“<%=”和“%>”符号之间不能插入语句只能插入表达式,而且“<%=”是一个完整的符号,其间不能有空格,“%>”符号也是如此。

3.4.2　技能操作

利用Java表达式在JSP页面中输出相应的内容。具体任务如下：

- 借助Math类中提供的方法,输出正弦值、输出立方值、输出平方根值；
- 输出算术表达式的值、输出逻辑表达式的值。

代码模板eg3_4.jsp如下：

```
<%@ page language="java" contenttype="text/html; charset=gbk"
    pageencoding="gbk"%>
<html>
<head>
    <title>java表达式</title>
</head>
<body bgcolor="cyan">
    <p>sin(3.14/6)等于
      <%=Math.sin(3.14/6)%>
    <p>8的立方是:
      <%=Math.pow(8,3)%>
    <p>196的平方根等于
      <%=Math.sqrt(196)%>
    <p>123456乘以2等于
      <%=123456*2%>
    <p>10000大于9999吗?答案:
```

```
        <%=10000>9999%>
    </body>
</html>
```

运行效果如图 3-9 所示。

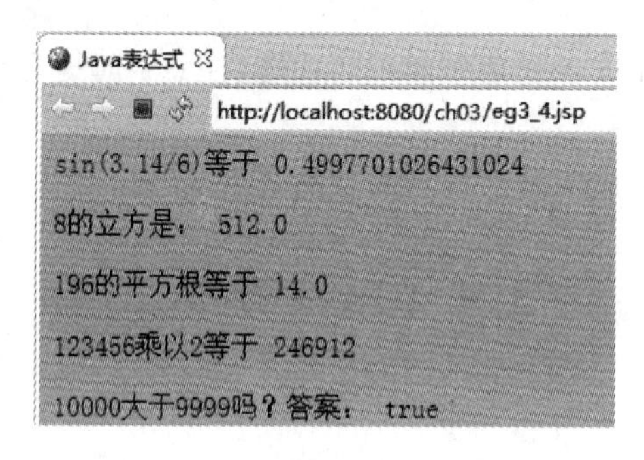

图 3-9 Java 表达式页面效果

3.4.3 拓展训练

1. 利用 Java 程序片和 Java 表达式，在 JSP 页面中输出一个 10 行 10 列的表格，每一个单元格中显示行、列坐标值，最终运行效果如图 3-10 所示。

http://localhost:8080/Test/

<table>
<tr><td>[1] | [1]</td><td>[1] | [2]</td><td>[1] | [3]</td><td>[1] | [4]</td><td>[1] | [5]</td><td>[1] | [6]</td><td>[1] | [7]</td><td>[1] | [8]</td><td>[1] | [9]</td><td>[1] | [10]</td></tr>
<tr><td>[2] | [1]</td><td>[2] | [2]</td><td>[2] | [3]</td><td>[2] | [4]</td><td>[2] | [5]</td><td>[2] | [6]</td><td>[2] | [7]</td><td>[2] | [8]</td><td>[2] | [9]</td><td>[2] | [10]</td></tr>
<tr><td>[3] | [1]</td><td>[3] | [2]</td><td>[3] | [3]</td><td>[3] | [4]</td><td>[3] | [5]</td><td>[3] | [6]</td><td>[3] | [7]</td><td>[3] | [8]</td><td>[3] | [9]</td><td>[3] | [10]</td></tr>
<tr><td>[4] | [1]</td><td>[4] | [2]</td><td>[4] | [3]</td><td>[4] | [4]</td><td>[4] | [5]</td><td>[4] | [6]</td><td>[4] | [7]</td><td>[4] | [8]</td><td>[4] | [9]</td><td>[4] | [10]</td></tr>
<tr><td>[5] | [1]</td><td>[5] | [2]</td><td>[5] | [3]</td><td>[5] | [4]</td><td>[5] | [5]</td><td>[5] | [6]</td><td>[5] | [7]</td><td>[5] | [8]</td><td>[5] | [9]</td><td>[5] | [10]</td></tr>
<tr><td>[6] | [1]</td><td>[6] | [2]</td><td>[6] | [3]</td><td>[6] | [4]</td><td>[6] | [5]</td><td>[6] | [6]</td><td>[6] | [7]</td><td>[6] | [8]</td><td>[6] | [9]</td><td>[6] | [10]</td></tr>
<tr><td>[7] | [1]</td><td>[7] | [2]</td><td>[7] | [3]</td><td>[7] | [4]</td><td>[7] | [5]</td><td>[7] | [6]</td><td>[7] | [7]</td><td>[7] | [8]</td><td>[7] | [9]</td><td>[7] | [10]</td></tr>
<tr><td>[8] | [1]</td><td>[8] | [2]</td><td>[8] | [3]</td><td>[8] | [4]</td><td>[8] | [5]</td><td>[8] | [6]</td><td>[8] | [7]</td><td>[8] | [8]</td><td>[8] | [9]</td><td>[8] | [10]</td></tr>
<tr><td>[9] | [1]</td><td>[9] | [2]</td><td>[9] | [3]</td><td>[9] | [4]</td><td>[9] | [5]</td><td>[9] | [6]</td><td>[9] | [7]</td><td>[9] | [8]</td><td>[9] | [9]</td><td>[9] | [10]</td></tr>
<tr><td>[10] | [1]</td><td>[10] | [2]</td><td>[10] | [3]</td><td>[10] | [4]</td><td>[10] | [5]</td><td>[10] | [6]</td><td>[10] | [7]</td><td>[10] | [8]</td><td>[10] | [9]</td><td>[10] | [10]</td></tr>
</table>

图 3-10 表格页面效果

要求编写、调试、运行下面给出的模板代码，体会 html 语言和 Java 程序片以及 Java 表达式相互之间的协作关系。

```
<%@ page language="java" import="java.util.*" pageEncoding="UTF-8"%>
<html>
    <head>
      <title>输出 10 行 10 列的表格</title>
    </head>
    <body>
```

```
        <table cellpadding="0" cellspacing="0" border="1" width="100%">
        <%
          int rows = 10;          // 多少行
          int cols = 10;         // 多少列
          for(int i = 0; i < rows; i++ ){
      %>
       <tr align="center" height="30">
      <%
          for(int j = 0; j < cols; j++ ){
      %>
        <td>[<%=i+1 %>] | [<%=j+1 %>]</td>
      <%
          }
        }
      %>
   </table>
     </body>
</html>
```

2. 综合运用变量的定义、Java程序片和Java表达式等所学内容，完成在JSP页面中输出九九乘法表的功能，最终效果如图3-11所示。

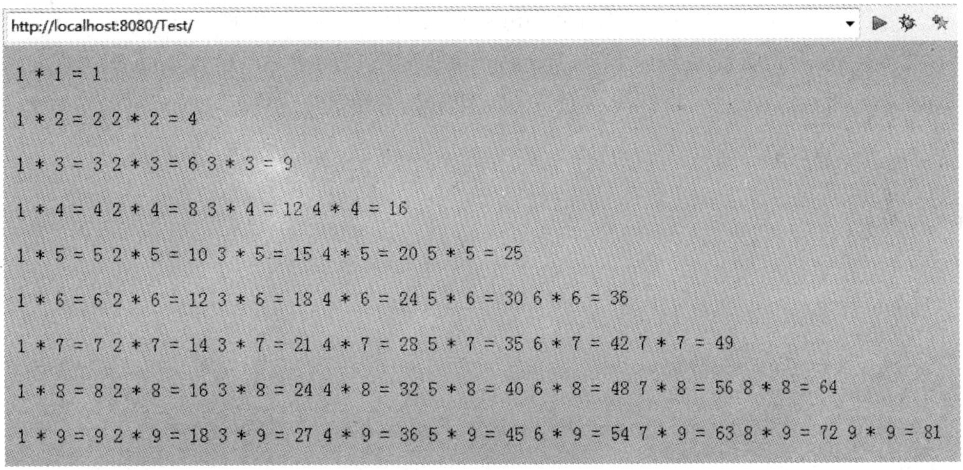

http://localhost:8080/Test/

1 * 1 = 1
1 * 2 = 2 2 * 2 = 4
1 * 3 = 3 2 * 3 = 6 3 * 3 = 9
1 * 4 = 4 2 * 4 = 8 3 * 4 = 12 4 * 4 = 16
1 * 5 = 5 2 * 5 = 10 3 * 5 = 15 4 * 5 = 20 5 * 5 = 25
1 * 6 = 6 2 * 6 = 12 3 * 6 = 18 4 * 6 = 24 5 * 6 = 30 6 * 6 = 36
1 * 7 = 7 2 * 7 = 14 3 * 7 = 21 4 * 7 = 28 5 * 7 = 35 6 * 7 = 42 7 * 7 = 49
1 * 8 = 8 2 * 8 = 16 3 * 8 = 24 4 * 8 = 32 5 * 8 = 40 6 * 8 = 48 7 * 8 = 56 8 * 8 = 64
1 * 9 = 9 2 * 9 = 18 3 * 9 = 27 4 * 9 = 36 5 * 9 = 45 6 * 9 = 54 7 * 9 = 63 8 * 9 = 72 9 * 9 = 81

图3-11　九九乘法表页面效果

要求将如下代码模板补充完整，并进行调试运行。

```
<%@ page contentType="text/html;charset=gb2312" %>
<html>
【代码段1】//定义整型变量i和j
<body bgcolor="gray">
  <%
```

```
      for(i=1;i<=9;i++){
          for(j=1;j<=i;j++){
  %>
  【代码段 2】//输出乘法表内容
    <%}%>
    <p>
  <%}%>
</body>
</html>
```

3.5 JSP 指令标记

为了更好地完成 Web 页面功能,JSP 引入了指令标记。本节将介绍常用的两个指令标记：page 指令标记和 include 指令标记。

3.5.1 知识要点

1. page 指令标记

page 指令标记用来定义 JSP 页面中的一些属性及其属性值。具体包括 import 属性、contentType 属性、pageEncoding 属性、language 属性、session 属性、isELIgnored 属性(只限 JSP 2.0)、buffer 属性、autoFlush 属性、info 属性、errorPage 属性、isErrorPage 属性、isThreadSafe 属性、extends 属性。在使用时,采用如下格式：

```
<%@page 属性 1="属性 1 的值"属性 2="属性 2 的值"… 属性 n="属性 n 的值" %>
```

或者

```
<%@page 属性 1="属性 1 的值" %>
<%@page 属性 2="属性 2 的值" %>
⋮
<%@page 属性 n="属性 n 的值" %>
```

在本书中只介绍 import、contentType、pageEncoding 三个常用属性。

1) import 属性

使用 page 指令标记的 import 属性可以为 JSP 页面指定应该引入的包及包中的类,以便 JSP 页面在变量和方法的定义、Java 程序片、Java 表达式中正常使用相应包中的类。

在同一个 JSP 文件中,page 指令标记的 import 属性可以多次出现,也就是说,可以指定多个 import 属性值,导入多个包中的多个类。例如,下面的指令：

```
<%@ page import="java.util.*, china.dalian.*" %>
```

表示 java.util 包和 china.dalian 包中的所有类在使用时无须再给出明确的包标识符。

另外,import 属性的书写位置没有特定要求,但通常情况下,习惯将 import 语句放在 JSP 文件顶部附近或是放在相应的包首次使用之前。

2) contentType 属性

使用 page 指令标记的 contentType 属性可以用来设置 contentType 响应报头,指明即将发送到客户端的文档的 MIME 类型(MIME 类型是设定某种文件用匹配的一种应用程序来打开的方式类型)和 JSP 页面字符的编码。例如,如果希望客户端浏览器启用 HTML 解析器来解析执行所接收到的信息,就可以按照如下方式设置 contentType 属性值:

```
<%@ page contentType="text/html;charset=GB2312" %>
```

如果希望客户端浏览器启用本地的 MS-Excel 应用程序来解析执行所接收的信息,则可以按照如下方式设置 contentType 属性值:

```
<%@ page contentType="application/vnd.ms-excel" %>
```

如果不使用 page 指令标记为 contentType 指定属性值,那么 contentType 属性就取默认值,即"text/html;charset=ISO-8859-1"。

与 import 属性不同,在 JSP 文件中,page 指令标记只能为 contentType 属性指定一个值。下面的用法是错误的:

```
<%@ page contentType="application/vnd.ms-excel" %>
<%@ page contentType="text/html;charset=GB2312" %>
```

可以利用 page 指令标记为 contentType 属性指定的 MIME 类型有 text/html(超文本标记语言文本)、text/plain(普通文本)、application/rtf(rtf 文本)、image/gif(gif 图形)、image/jpeg(jpeg 图形)、audio/basic(au 声音文件)、audio/midi(midi 音乐文件)、audio/x-pn-realaudio(realaudio 音乐文件)、video/mpeg(mpeg 文件)、video/x-msvideo(avi 文件)、application/x-gzip(gzip 文件)、application/x-tar(tar 文件)、application/vnd.ms-excel(excel 文件)、application/msword(word 文件)。

3) pageEncoding 属性

如前所述,如果希望同时设置所接收到信息的 MIME 类型和 JSP 页面字符编码,可以使用 contentType 属性来完成。但是,如果只想更改字符集,使用 pageEncoding 属性更为简单。例如,中文 JSP 页面可以使用下面的语句:

```
<%@ page pageEncoding="GBK" %>
```

来设置。

关于前面提到的字符集,经常使用的有如下几种:

(1) ASCII。

ASCII(American Standard Code for Information Interchange,美国信息互换标准代

码),是基于常用英文字符的一套编码。

(2) ISO-8859-1。

ISO-8859-1 编码通常叫做 Latin-1,除收录 ASCII 字符外,还增加了其他一些语言和地区需要的字符。该编码是 Tomcat 服务器默认采用的字符编码。

(3) GB 2312。

GB 2312 码是中华人民共和国国家标准汉字信息交换用编码,简称国标码,是由国家标准总局发布的关于汉字的编码,通行于中国大陆和新加坡。

(4) GBK。

GBK 编码规范,除了完全兼容 GB 2312,还对繁体中文和一些不常用的字符进行了编码。GBK 是现阶段 Windows 和其他一些中文操作系统的默认字符集。

(5) Unicode。

Unicode 为统一的字符编码标准集,为地球上几乎所有地区每种语言中的每个字符设定了统一并且唯一的编码,以满足跨语言、跨平台进行文本转换、处理的要求。

(6) UTF-8。

UTF-8 是 Unicode 的一种变长字符编码。用在网页上可以在同一页面显示中文和其他语言。当处理包含多国文字的信息页面时一般选择用 UTF-8。

2. include 指令标记

网站中的多个 JSP 页面有时需要显示同样的信息,如该网站的 Logo 或是导航等,为了便于网站的管理与维护,通常在这些 JSP 页面的适当位置嵌入一个相同的文件以完成相应的功能。include 指令标记的用途就在于此,可以借助该标记在当前 JSP 页面中整体嵌入另一个文件。语法格式为:

```
<%@ include file="文件的地址" %>
```

其中被嵌入的文件必须是可访问和可使用的。

3.5.2 技能操作

使用 JSP 指令标记,完成相应的功能。具体任务如下:

- 利用 page 指令标记的 import 属性,导入 java.util 包中的类,完成在当前页面显示系统时间的功能。
- 利用 page 指令标记的 contentType 属性,指定客户端浏览器启用本地的 Microsoft Word 应用程序来解析执行所接收的信息。
- 利用 include 指令标记,在同一网站的不同页面顶端显示相同的 Logo 图片。

1. 利用 page 指令标记的 import 属性,导入 java.util 包中的类,完成在当前页面显示系统时间的功能。

代码模板 eg3_5.jsp 如下:

```
<%@ page language="java" contentType="text/html; charset=GBK"%>
<%@ page import="java.util.*"%>
<html>
<head>
    <title>JSP 指令标记</title>
</head>
<%!
 String s=null ;
%>
<body bgcolor="cyan">
    当前的日期和时间是:
    <%
     Date date=new Date();
     s = date.toString();
    %>
    <%= s%>
</body>
</html>
```

运行效果如图 3-12 所示。

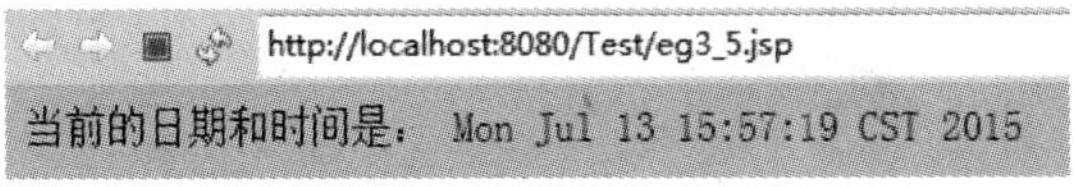

图 3-12　显示日期时间页面效果

2. 利用 page 指令标记的 contentType 属性，指定客户端浏览器启用本地的 Microsoft Word 应用程序来解析执行所接收的信息。

代码模板 eg3_6.jsp 如下：

```
<%@ page language="java" contentType="application/msword; charset=GBK"%>
<html>
<head>
    <title>JSP 指令标记</title>
</head>
<body bgcolor="cyan">
  <P>启用 Microsoft Word 应用程序处理所接收到的信息.
  <input type="text" size="12">
</body>
</html>
```

运行上述页面时会弹出如图 3-13 所示的“文件下载”对话框，单击“保存”按钮弹出如图 3-14 所示的对话框，选择相应的路径保存后，就会启用本地的 Microsoft Word 应用程序来显示当前页面内容，如图 3-15 所示。

图 3-13 “文件下载”对话框

图 3-14 选择文件保存路径

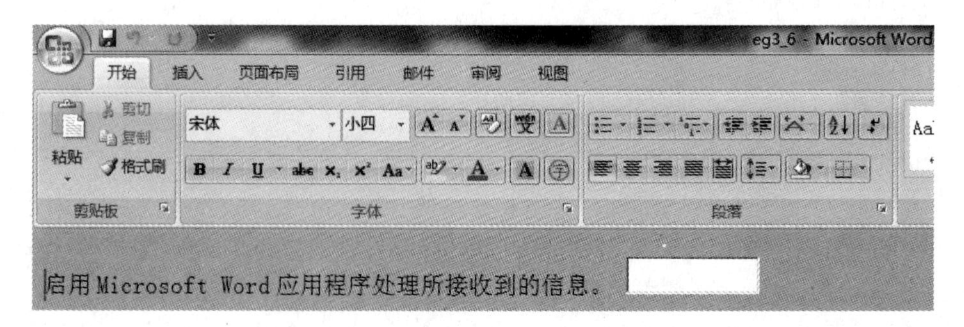

图 3-15 Microsoft Word 解析页面内容

3. 利用 include 指令标记，在同一网站的不同页面顶端显示相同的 logo 图片。

代码模板 eg3_7.jsp 如下：

```
<%@ page language="java" contentType="text/html; charset=GBK"%>
<html>
<head>
    <title>Logo 图片</title>
</head>
<body>
    <img src=" logo.jpg" >
</body>
</html>
```

代码模板 eg3_7_1.jsp 如下：

```
<%@ page language="java" contentType="text/html; charset=GBK" %>
<%@ include file="eg3_7.jsp" %>
<html>
<head>
    <title>include 指令标记</title>
</head>
<body>
  <p>
  <font size="20" color="red" >
  立警为公　执法为民
  </font>
</body>
</html>
```

运行上述代码，显示效果如图 3-16 所示。其中上半部分的图片来自于被包含进来的 eg3_7.jsp 页面效果，下半部分的文字“立警为公　执法为民”来自于 eg3_7_1.jsp 本身页面效果。

图 3-16　include 指令标记应用之一

代码模板 eg3_7_2.jsp 如下：

```
<%@ page language="java" contentType="text/html; charset=GBK" %>
<%@ include file="logo.jsp" %>
<html>
<head>
    <title>include 指令标记</title>
</head>
<body>
  <p>警察(武警或人民警察)
  <p>中国的警察包括武警和人民警察两大类。
  <p>“公安”广义上是指人民警察,分为公安警察、国家安全警察及司法警察。
  <p>人民警察是国家公务员,实行警监、警督、警司、警员的警衔制度,服装以藏黑为主色调。
  <p>武警全称中国人民武装警察部队,是中华人民共和国武装力量的一部分,是担负国家赋予的
安全保卫任务的部队。
</body>
</html>
```

运行上述代码,显示效果如图 3-17 所示。

图 3-17　include 指令标记应用之二

与前面的应用类似,图 3-17 中上半部分的图片来自于被包含进来的 eg3_7.jsp 页面效果,下半部分的文字来自于 eg3_7_2.jsp 本身页面效果。

3.5.3 拓展训练

1. 编写一个 JSP 页面，利用 page 指令标记的 import 属性将 java. util 包中的类、java. io 包中的类导入，然后根据实际需求进行使用。调试运行程序后观察页面效果。

2. 把 3.5.2 节任务 2 中 eg3_6. jsp 页面的 contentType 属性值修改为 application/vnd . ms-powerpoint，指定客户端浏览器启用本地的 Microsoft PowerPoint 应用程序来解析执行所接收的信息。运行修改后的页面观察页面效果。

3. 使用"记事本"编写一个文本文件 included. txt。included. txt 的每行有若干个英文单词，这些单词之间用空格分隔，每行之间用
分隔，具体如下：

included. txt 代码：

```
packag apple void back public
<br>
private throw class hello welcome
```

编写三个 JSP 页面：first. jsp、second. jsp 和 third. jsp，要求每个页面都包含 included . txt 文件，其余页面内容自行设计。调试运行程序后观察页面效果。

3.6 JSP 动作标记

动作标记是一种特殊的标记，它们将影响 JSP 运行时的功能。JSP 有 7 个动作标记，分别是：

jsp:include——用于动态引入一个 JSP 页面。

jsp:param——用于传递参数，必须与其他支持参数的标签一起使用。

jsp:forward——执行页面转向，将请求的处理转发到下一个页面。

jsp:plugin——用于下载 JavaBean 或 Applet 到客户端执行。

jsp:useBean——使用 JavaBean。

jsp:setProperty——修改 JavaBean 实例的属性值。

jsp:getProperty——获取 JavaBean 实例的属性值。

本节着重介绍前 4 个动作标记，其余的动作标记将在第 5 章详细介绍。

3.6.1 知识要点

1. include 动作标记

include 动作标记用于在当前 JSP 页面中动态包含另一个文件，即将当前 JSP 页面、被包含的文件各自独立地翻译为字节码文件。当前 JSP 页面执行到该动作标记时，才加载执行被包含文件的字节码。语法格式如下：

```
<jsp:include page="文件的地址"/>
```

或是

```
<jsp:include 动作标记 page= "文件的地址">
param 子标记
</jsp:include>
```

需要注意的是，当 include 动作标记不需要子标记时，必须使用第一种形式。在 3.5 节中介绍的指令标记与本节的 include 动作标记功能相似，都是将另一个文件嵌入到当前 JSP 页面中，但是其处理 include 指令标记方式和处理时机不同。include 指令标记是静态嵌入，即在编译阶段就处理嵌入的文件，被嵌入的文件在语法上和逻辑上依赖于当前 JSP 页面，其优点是执行速度快；而 include 动作标记是动态嵌入，即在 JSP 页面运行时才处理嵌入的文件，被嵌入的文件在语法上和逻辑上相对独立，其优点是可以借助 param 子标记灵活处理所包含的文件，与此同时也会造成执行速度减慢。

2. param 动作标记

param 动作标记用于设置参数值，这个动作标记本身不能独立使用，独立的 param 动作标记没有实际意义，需作为 jsp:include、jsp:forward、jsp:plugin 标记的子标记来使用，以完成相应的功能。例如当该标记与 jsp:include 动作标记一起使用时，可以将 param 标记中的值传递到 include 动作标记要加载的文件中，被加载的 JSP 文件可以使用 Web 服务器提供的 request 内置对象获取 include 动作标记的 param 子标记中属性所提供的值。

语法格式如下：

```
<jsp:param name="paramName" value="paramValue">
```

3. forward 动作标记

forward 动作标记用于将页面响应转发给另外的页面，既可以转发给静态的 HTML 页面，也可以转发到动态的 JSP 页面。换言之，就是从该动作标记处停止执行当前 JSP 页面，而转向指定的 HTML 页面或 JSP 页面去执行。语法格式如下：

```
<jsp:forward page="将转向的页面"/>
```

或是

```
<jsp:forward page="将转向的页面">
    param 子标记
</jsp:forward>
```

需要注意的是，当 forward 动作标记不需要子标记时，必须使用第一种形式。

4. plugin 动作标记

plugin 动作标记用于执行一个 applet(Java 小应用程序)，因为有些浏览器并不支持 Java applet 程序，所以可以借助 plugin 动作标记保证客户端浏览器能够顺利执行 Java applet程序。在具体执行过程中，plugin动作标记指示JSP页面加载Java plugin，该插件由客户负责下载，并使用该插件来运行Java applet程序。语法格式如下：

```
<jsp:plugin
    type="applet "
    code="字节码文件"
    codebase="字节码文件路径"
    [ height="小程序高度值" ]
    [ width="小程序宽度值" ]
    [ jreversion="JRE版本号" ]
    [ <jsp:fallback>提示用户的文本信息</jsp:fallback> ]
</jsp:plugin>
```

下面的例子代码，表明小应用程序的字节码文件是A.class，且该字节码文件存放在当前Web服务目录中；小应用程序显示高度为180像素，宽度为200像素，JRE版本号为1.1；当插件不能启动时显示给用户一段提示性文字"Unable to load applet"。

```
<jsp:plugin
    type="applet "
    code="A.class"
    codebase="."
    height="180"
    width="200"
    jreversion="1.1"
    <jsp:fallback>Unable to load applet.</jsp:fallback>
</jsp:plugin>
```

需要注意的是，上面语法格式中被"[]"括起来的部分为可选项，可以根据实际情况来决定是否对其进行设置。标记<jsp:fallback>是<jsp:plugin>动作标记的一部分，并且只能在<jsp:plugin>动作中使用。

3.6.2 技能操作

使用JSP动作标记，完成相应的功能。具体任务如下：

- 利用include动作标记和param动作标记完成简单的欢迎登录页面。
- 利用forward动作标记和param动作标记实现奇偶页的不同跳转页面。

1. 利用include动作标记和param动作标记完成简单的欢迎登录页面。

代码模板eg3_8.jsp如下：

```
<%@ page language="java" contentType="text/html; charset=GB2312" %>
<html>
    <head>
        <title>JSP动作标记</title>
    </head>
```

```
        <body bgcolor="cyan">
        <%
        String user="sa",password="123456";
        %>
        <jsp:include page="success.jsp">
          <jsp:param name="text1" value="<%=user %>"/>
          <jsp:param name="text2" value="<%=password %>"/>
        </jsp:include>
      </body>
    </html>
```

代码模板 success.jsp 如下：

```
<%@ page language="java" contentType="text/html; charset=GBK"%>
登录成功,欢迎
<br>
您的用户名是:
<%=request.getParameter("text1")%>
<br>
您的密码是:
<%=request.getParameter("text2")%>
```

最终运行结果如图 3-18 所示。可见,eg3_8.jsp 中通过下面两行语句:

```
<jsp:param name="text1" value="<%=user %>"/>
<jsp:param name="text2" value="<%=password %>"/>
```

设定了传递的参数列表,即名称为 text1 的参数其值为 user 变量的值,也就是"sa"字符串;名称为 text2 的参数其值为 password 变量的值,也就是"123456"字符串。

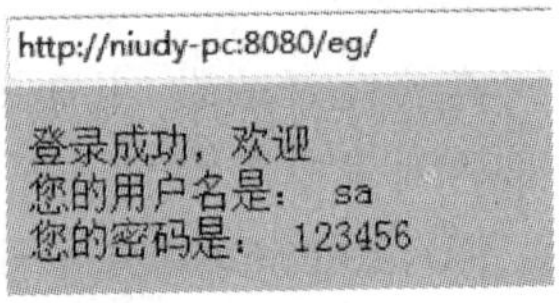

http://niudy-pc:8080/eg/

登录成功，欢迎
您的用户名是： sa
您的密码是： 123456

图 3-18　欢迎登录页面效果

2. 利用 forward 动作标记和 param 动作标记实现奇偶页的不同跳转页面。

代码模板 eg3_9.jsp 如下：

```
<%@ page language="java" contentType="text/html; charset=GBK"%>
<html>
  <body>
      <%
      int number=(int)(Math.random() * 100+1);
      if(number%2!=0){
      %>
      <jsp:forward page="eg3_9_1.jsp">
      <jsp:param name="text" value="<%=number%>"/>
      </jsp:forward>
      <%
```

```
        }
        else
        {
        %>
        <jsp:forward page="eg3_9_2.jsp">
        <jsp:param name="text" value="<%=number%>"/>
        </jsp:forward>
        <% }
        %>
    </body>
</html>
```

代码模板 eg3_9_1.jsp 如下：

```
<%@ page language="java" contentType="text/html; charset=GBK"%>
<html>
<body bgcolor="cyan">
传递过来的随机数为:
<%=request.getParameter("text")%>
<p>
该奇数页图片为:
<p>
<img src="D:\picture1.jpg">
</body>
```

代码模板 eg3_9_2.jsp 如下：

```
<%@ page language="java" contentType="text/html; charset=GBK"%>
<html>
<body bgcolor="cyan">
传递过来的随机数为:
<%=request.getParameter("text")%>
<p>
该偶数页图片为:
<p>
<img src="D:\picture2.jpg">
</body>
</html>
```

具体运行效果如图 3-19 和图 3-20 所示。从图中可以看出，当 eg3_9.jsp 中产生的随机数为奇数时，则跳转到 eg3_9_1.jsp 页面，显示图像为人物左脸；当 eg3_9.jsp 中产生的随机数为偶数时，则跳转到 eg3_9_2.jsp 页面，显示图像为人物右脸。

图 3-19 奇数页跳转页面效果

图 3-20 偶数页跳转页面效果

3.6.3 拓展训练

1. 编辑下面给出的代码模板，调试运行并观察页面效果，体会 include 指令标记和 include 动作标记的联系与区别。

代码模板 static.html 如下

```
<%@ page language="java" contentType="text/html; charset=GBK"%>
<html>
    <head>
        <title>JSP 动作标记</title>
    </head>
    <body bgcolor="cyan">
        name:<input type="text" value="123">
        password:<input type="text" value="123">
        <input type="button" value="login" name="button">
    </body>
</html>
```

代码模板 dynamic.jsp 如下：

```
<%@ page language="java" contentType="text/html; charset=GBK"%>
<html>
    <head>
        <title>JSP 动作标记</title>
    </head>
    <body>
```

```
        我来自第 2 个文件.
    </body>
</html>
```

代码模板 index.jsp 如下：

```
<%@ page language="java" contentType="text/html; charset=GBK"%>
<html>
    <head>
        <title>JSP 动作标记</title>
    </head>
    <body bgcolor="cyan">
        这是 include 指令标记举例.
        <br><%@ include file="static.html" %>
        <br>这是 include 动作标记举例.
        <br><jsp:include page="dynamic.jsp" flush="true"/>
    </body>
</html>
```

2. 模仿 3.6.2 节中的例题，使用 forward 动作标记和 param 动作标记实现页面跳转功能。要求：编写一个静态 HTML 文件 decide.html，在该文件中定义一个变量 i 并为其赋初值，当 i 的值大于 0 时，跳转到页面 positive.jsp；否则跳转到页面 negative.jsp。positive.jsp 和 positive.jsp 页面内容可以自行设计。

3.7 小　　结

- 一个 JSP 页面通常由普通的 HTML 标记、JSP 注释、Java 动态元素（包括变量和方法的声明、Java 程序片、Java 表达式）以及 JSP 标记（包括指令标记、动作标记和自定义标记）组成。
- 在一个 Java 程序片中声明的变量称为 JSP 页面的局部变量，它们在 JSP 页面后续的所有程序片部分以及表达式部分有效。一个用户对 JSP 页面局部变量操作的结果，不会影响到其他用户。
- JSP 页面成员变量是被所有用户共享的变量，任何用户对 JSP 页面成员变量操作的结果，都会影响其他用户。
- page 指令标记用来定义整个 JSP 页面的一些属性以及这些属性的值。page 指令只能为 contentType 属性指定一个值，但可以为 import 属性指定多个值。
- include 指令标记是先将当前 JSP 页面与要嵌入的文件合并成一个新的 JSP 页面，然后再由 JSP 引擎将新页面转化为 Java 文件处理并运行。而 include 动作标记在把 JSP 页面转译成 Java 文件时，并不合并两个页面；而是在 Java 文件的字节码文件被加载和执行时，才去处理 include 动作标记中引入的文件。

第 4 章

JSP 内置对象

为简化 Web 页面的开发,JSP 本身提供了一些内置对象。这些内置对象在 JSP 页面中不需要声明和实例化,就可以直接在 Java 程序片和 Java 表达式中使用,它们由 Web 服务器负责实现和管理。

JSP 提供的内置对象有 request 对象、response 对象、session 对象、application 对象、out 对象、pageContext 对象、config 对象、exception 对象。本章将主要介绍前 4 个常用内置对象的使用方法,其他内置对象及其常用方法可参考本书附录 C。

4.1 request 对象

4.1.1 知识要点

request 对象代表请求对象,它是实现了 javax. servlet. ServletRequest 接口的一个实例。当用户请求一个 JSP 页面时,JSP 页面所在的服务器将用户发出的所有请求信息封装在内置对象 request 中,使用该对象就可以获取用户提交的信息。

下面就着重学习 request 对象的两个常用方法,即 getParameter(String name)方法和 getParameterValues(String name)方法,它们的主要功能是用来获取客户提交信息。

1. public String getParameter(String name)方法

该方法以字符串的形式返回客户提交的信息,而客户向 Web 服务器提交信息的方式通常是使用 HTML 表单来进行的,其一般格式为:

```
<form method="get"|"post"action="提交信息将要传送到的目标地址">
    提交手段(包括文本框、文本区、下拉列表、单选按钮、复选框等)
</form>
```

其中<form>是表单标记,method 取值 get 或 post。get 方法和 post 方法的主要区别是:使用前者提交的信息会在提交过程中显示在浏览器地址栏中,而使用后者提交的信息不会显示在地址栏中。例如:

```
<form method="post"action="destination.jsp">
    <input type="text" name="loginName" value="sa"/>
    <input type="submit" name= "button"value="注册"/>
</form>
```

单击“注册”按钮后，JSP页面destination.jsp就可以借助下面的代码：

```
String nameInfo = request.getParameter("loginName");
```

来获得客户端提交的文本框中预先设定的信息“sa”，并进行后续的处理和操作，且提交的信息不会显示在浏览器地址栏中。

2. public String[] getParameterValues(String name)

该方法以字符串数组的形式返回客户端向服务器端传递的指定参数名的所有值。例如：

```
<form method="post"action="destination.jsp">
    您的兴趣爱好:
        <input type="checkbox" name="hobbies" value="reading"/>阅读
        <input type="checkbox" name="hobbies" value="music"/>音乐
        <input type="checkbox" name="hobbies" value="surfing"/>上网
        <input type="checkbox" name="hobbies" value="swimming"/>游泳
        <input type="checkbox" name="hobbies" value=" badminton"/>羽毛球
</form>
```

如果用户选择了“阅读”、“音乐”、“上网”和“游泳”四项兴趣爱好，那么提交表单后JSP页面destination.jsp就可以借助request内置对象的getParameterValues方法获取用户提交的信息，即复选框hobbies的属性值reading、music、surfing和swimming，并将信息存放到一个数组中，该数组的元素的值与复选框的属性值一一对应。具体代码如下：

```
String hobbies[] = request.getParameterValues("hobbies");
```

4.1.2 技能操作

使用request内置对象获取客户提交的信息。具体任务如下：

编写两个JSP页面eg4_1.jsp和destination.jsp，eg4_1.jsp通过表单向destination.jsp提交输入的用户名和选择的兴趣爱好，destination.jsp负责获得表单中提交的信息并显示。

代码模板eg4_1.jsp如下：

```
<%@ page language = "java" contentType = " text/html; charset = GBK" pageEncoding =
"GBK"%>
<html>
<head>
<title>eg4_1.jsp</title>
</head>
<body bgcolor="cyan">
    <form action="destination.jsp" method="post">
        用户名: <input type="text" name="loginName"/>
        <br>
```

```
            您的兴趣爱好:
            <input type="checkbox" name="hobbies" value="reading"/>阅读
            <input type="checkbox" name="hobbies" value="music"/>音乐
            <input type="checkbox" name="hobbies" value="surfing"/>上网
            <input type="checkbox" name="hobbies" value="swimming"/>游泳
            <input type="checkbox" name="hobbies" value=" badminton"/>羽毛球
            <br>
            <input type="submit" value="注册"/>
        </form>
</body>
</html>
```

代码模板 destination.jsp 如下:

```
<%@ page language = "java" contentType = "text/html; charset = GBK" pageEncoding =
"GBK"%>
<html>
<head>
<title>getValue.jsp</title>
</head>
<body bgcolor="cyan">
<%
    String nameInfo= request.getParameter("loginName"); //获得 eg4_1.jsp 页面中输入的姓名
    String hobbies[]= request.getParameterValues("hobbies"); //获得 eg4_1.jsp 页面中选择的兴趣
%>
    您输入的姓名是: <%=nameInfo %><br>
    您的兴趣爱好有:
    <%
        for(int i=0;i<hobbies.length;i++){
                out.print(hobbies[i]+" ");}
    %>
</body>
</html>
```

最终运行页面效果如图 4-1 和图 4-2 所示。

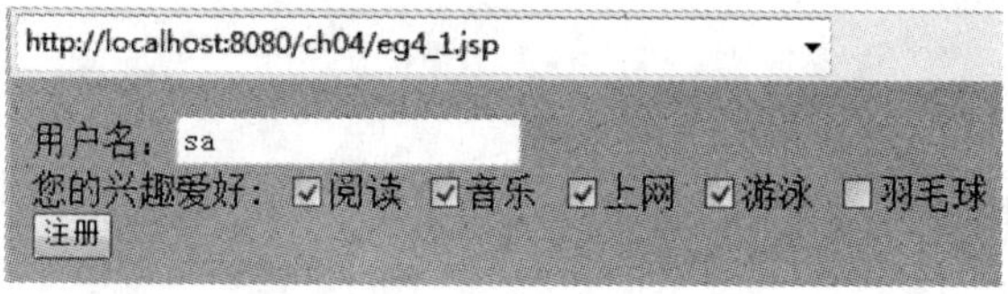

图 4-1　例子 4_1 页面效果之一

图4-2　例子4_1页面效果之二

4.1.3　拓展训练

1. 如果在上述eg4_1.jsp页面中不选择任何兴趣爱好，而直接单击“注册”按钮，观察destination.jsp页面中会出现什么结果？在你的浏览器中是否出现如图4-3所示的信息，即NullPointerException（空指针）异常。要求：按照下面的提示解决eg4_1.jsp中的NullPointerException（空指针）异常问题。

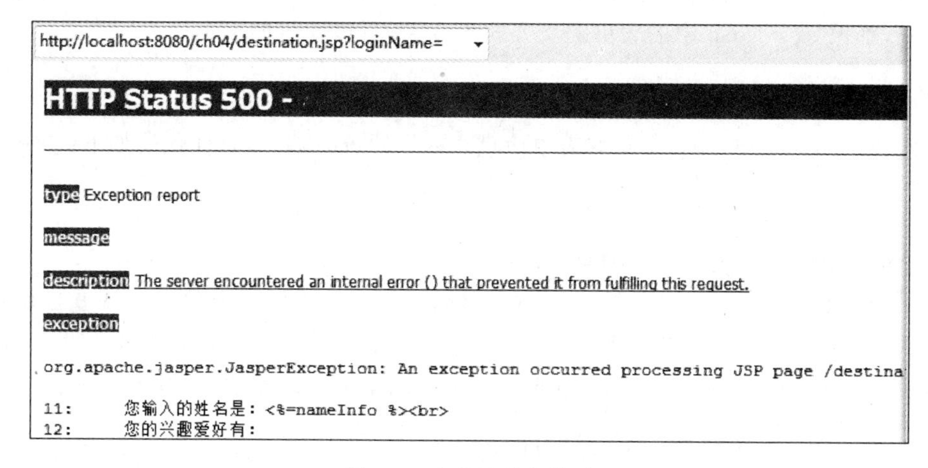

图4-3　空指针异常提示

为了避免在运行时出现NullPointerException异常，可以在destination.jsp页面中显示兴趣爱好部分代码修改为：

```
if(hobbies!=null){
    for(int i=0;i<hobbies.length;i++){
        out.print(hobbies[i]+" ");
    }
}
```

如此修改后，即使用户没有对兴趣爱好进行任何选择，也不会出现上面提到的空指针异常了，最终的显示效果如图4-4所示。这样操作，有效保证了JSP页面良好的交互性。

2. 如果在eg4_1.jsp页面的文本框中输入中文姓名，观察destination.jsp页面中获得的用户名是什么效果？在你的浏览器中是否出现如图4-5所示的乱码（用户名由“?”等组成）。要求：按照下面的提示解决eg4_1.jsp中的中文乱码问题。

http://localhost:8080/ch04/destination.jsp?loginName=

您输入的姓名是：
您的兴趣爱好有：

图 4-4　修改后的例子 4_1 页面效果

http://localhost:8080/ch04/destination.jsp

您输入的姓名是：???ì??
您的兴趣爱好有：　reading swimming

图 4-5　中文乱码

解决乱码常用的方法有两种：

（1）使用 setCharacterEncoding(String code)设置统一字符编码。

request 对象提供了方法 setCharacterEncoding(String code)设置编码，其中参数 code 以字符串形式传入要设置的编码格式，但这种方法仅对于提交方式是 post 的表单有效（表单默认的提交方式是 get）。例如，我们使用该方法解决 eg4_1.jsp 和 destination.jsp 页面中出现的中文乱码问题，需要完成两件事：

首先，确保 eg4_1.jsp 中的表单提交方式为 post，具体代码如下：

```
<form action="destination.jsp" method="post">
```

其次，在 destination.jsp 中获取表单信息之前设置统一编码，具体代码如下：

```
request.setCharacterEncoding("GBK");
```

（2）对获取的信息进行重新编码。

通过内置对象 request 获取到字符串的值后，对该字符串使用 ISO-8859-1 重新编码，并把编码的结果存放到一个字节数组中，然后再把这个字节数组转化为字符串。例如，使用该方法解决 eg4_1.jsp 和 destination.jsp 页面中出现的中文乱码问题，具体代码如下：

```
String name=request.getParameter("loginName");
byte b[]=name.getBytes("ISO-8859-1");
name=new String(b);
```

4.2　response 对象

4.2.1　知识要点

当用户请求服务器的一个页面时，会提交一个 HTTP 请求，服务器收到请求后，返回 HTTP 响应。4.1 节中已经介绍过，request 对象能够将请求信息进行封装，而本节中将要学习的 response 对象能够为用户的请求做出动态响应。可以说，request 对象与 response 对象是相互匹配的两个对象。下面就是 response 对象做出动态响应的三个方面的内容。

1. 动态响应 contentType 属性

JSP 页面用 page 指令标记设置了页面的 contentType 属性值，response 对象就按照这种属性值的方式对客户做出响应。第3章已经介绍过，在page指令中只能为contentType属性指定一个值，所以如果想动态指定contentType属性值，换一种方式来响应客户，可以让response对象调用setContentType(String s)方法来重新设置网页响应的MIME类型。常见的MIME类型有text/html（超文本标记语言文本）、text/plain（普通文本）、application/rtf(rtf文本)、image/gif(gif图形)、image/jpeg(jpeg图形)、audio/basic(au声音文件)、audio/midi(midi音乐文件)、audio/x-pn-realaudio(realaudio音乐文件)、video/mpeg(mpeg文件)、video/x-msvideo(avi文件)、application/x-gzip(gzip文件)、application/x-tar(tar文件)、application/vnd.ms-excel(excel文件)、application/msword(word文件)等。

例如，下面的语句：

```
response.setContentType("application/msword;charset=GB2312");
```

将会提示用户启用Microsoft Word来显示或保存当前页面。

2. 设置响应表头(HTTP文件头)

response对象可以通过方法setHeader(String name,String value)设置指定名字的HTTP文件头的值，以此来操作HTTP文件头。response对象设置的新值将会覆盖原值。如果希望某页面每5秒钟刷新一次，那么应在该页面中添加如下代码：

```
response.setHeader("refresh","5");
```

3. response重定向

在需要将用户引导至另一个页面时，可以使用reponse对象的sendRedirect (String url)方法实现用户的重定向。例如用户输入的表单信息不完整，应该再次被重定向到输入页面。例如下面的语句，会重定向到input.jsp页面：

```
response.redirect("input.jsp");
```

4.2.2 技能操作

使用response内置对象动态响应客户的请求，具体任务如下：

- 动态响应contentType属性值。
- 设置响应表头(HTTP文件头)。
- 重定向到指定页面。

1. 动态响应contentType属性值

编写JSP页面eg4_2.jsp，客户通过单击页面上的不同按钮，可以改变页面响应的MIME类型。当单击Microsoft Word按钮时，JSP页面动态地改变contentType的属性值为application/msword，客户浏览器启用本地的Word软件来显示当前页面内容；当单击Microsoft Excel按钮时，JSP页面动态地改变contentType的属性值为application/vnd.ms-excel，客户浏览器启用本地的Excel软件来显示当前页面内容；当单击Microsoft

PowerPoint 按钮时,JSP 页面动态地改变 contentType 的属性值为 application/msword,客户浏览器启用本地的 PowerPoint 软件来显示当前页面内容。

代码模板 eg4_2.jsp 如下：

```
<%@ page language = " java"  contentType = " text/html;  charset = GBK "  pageEncoding =
"GBK"%>
<html>
<head>
<title>eg4_2.jsp</title>
</head>
<body bgcolor="cyan">
     <form action=""method="post">
    <p>使用 response 动态响应 contentType 属性值:
    <p>单击 Microsoft Word 按钮,使用 Word 显示页面内容
    <input type="submit" value="Word" name="submit">
    <p>单击 Microsoft Excel 按钮,使用 Excel 显示页面内容
    <input type="submit" value="Excel" name="submit">
    <p>单击 Microsoft PowerPoint 按钮,使用 PowerPoint 显示页面内容
     <input type="submit" value="PowerPoint" name="submit">
          <%
          String str = request.getParameter("submit");
          if ("Word".equals(str)) {
          //response 调用 setContentType 方法设置 MIME 类型为 application/msword
          response.setContentType("application/msword");
          } elseif ("Excel".equals(str)) {
          //response 调用 setContentType 方法设置 MIME 类型为 application/ vnd.ms-excel
          response.setContentType("application/vnd.ms-excel");}
          %>
     </form>
</body>
</html>
```

最终运行页面效果如图 4-6～图 4-8 所示。

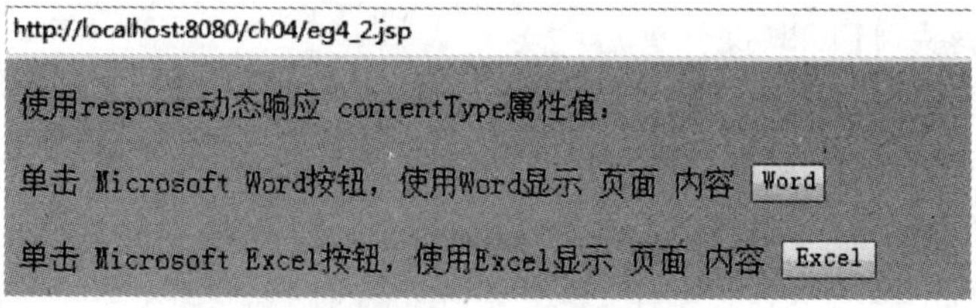

http://localhost:8080/ch04/eg4_2.jsp

使用response动态响应 contentType属性值：

单击 Microsoft Word按钮，使用Word显示 页面 内容 Word

单击 Microsoft Excel按钮，使用Excel显示 页面 内容 Excel

图 4-6 例子 4_2 页面效果

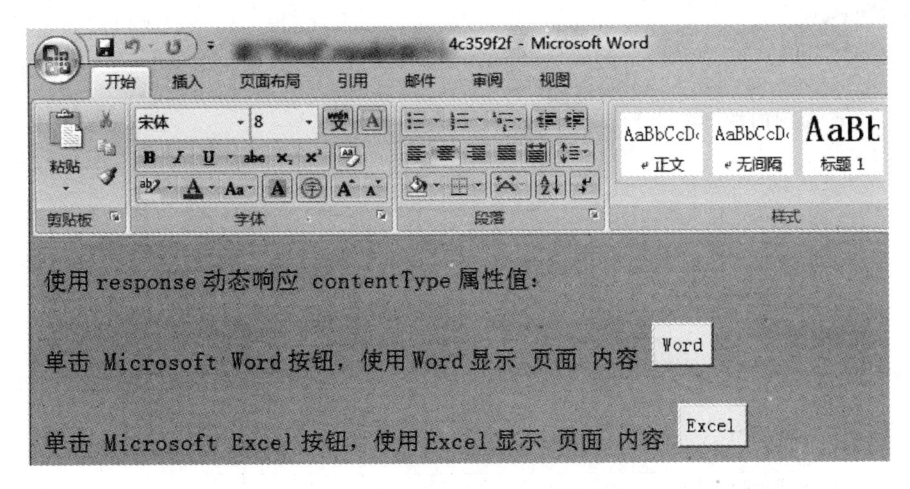

图 4-7　使用 Word 显示页面内容

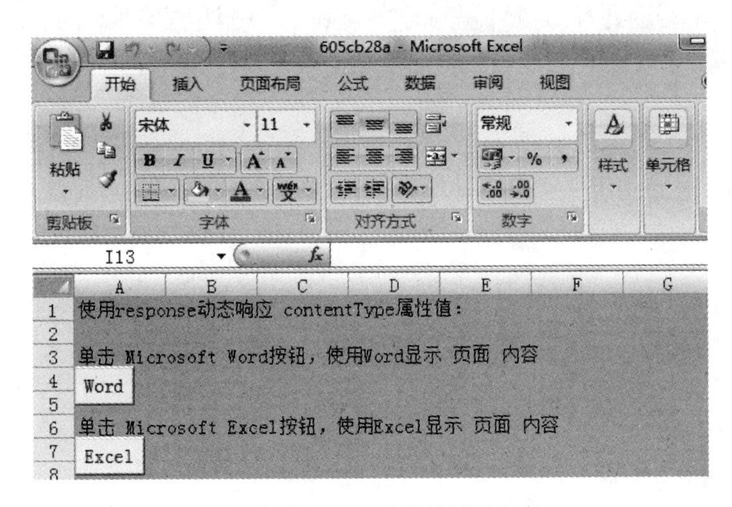

图 4-8　使用 Excel 显示页面内容

2. 设置响应表头(HTTP 文件头)。

编写一个 JSP 页面 eg4_3.jsp，在该页面中使用 response 对象设置一个响应头 refresh，其值是 5。那么用户收到这个头之后，该页面会每 5 秒钟刷新一次。

代码模板 eg4_3.jsp 如下：

```
<%@ page language = "java" contentType = "text/html; charset = GBK" pageEncoding =
"GBK"%>
<%@ page import="java.util.*" %>
<html>
<head>
<title>eg4_3.jsp</title>
```

```
</head>
<body>
    <h2>该 JSP 页面每隔 5 秒钟就会刷新 1 次,请注意观察时间变化</h2>
    <p>当前的秒钟时间是:
    <%
        Date d=new Date();
        out.print(""+d.getSeconds());
        response.setHeader("refresh","5");
        //使用 response 对象设置一个响应头"refresh",其值是"3".
    %>
  </body>
  </html>
```

最终运行页面效果如图 4-9 所示。

http://localhost:8080/ch04/eg4_3.jsp

该JSP页面每隔5秒钟就会 刷新1次，请注意观察 时间变化

当前的秒钟时间是： 1

图 4-9　例子 4_3 页面效果之一

当 5 秒钟过后,该 JSP 页面会自动刷新,出现如图 4-10 所示的页面效果。

http://localhost:8080/ch04/eg4_3.jsp

该JSP页面每隔5秒钟就会 刷新1次，请注意观察 时间变化

当前的秒钟时间是： 6

图 4-10　例子 4_3 页面效果之二

3. 重定向到指定页面

编写两个 JSP 页面,即 eg4_4.jsp 和 verify.jsp,如果在页面 eg4_4.jsp 中输入正确的管理员账号 sa 以及管理员密码 admin123,单击"登录"按钮后将账号和密码信息提交给页面 verify.jsp；如果输入不正确,则重新定向到 eg4_4.jsp 页面。

代码模板 eg4_4.jsp 如下：

```
<%@ page language = "java" contentType = "text/html; charset = GBK" pageEncoding =
"GBK"%>
<html>
<head>
 <title>eg4_4.jsp</title>
</head>
<body bgcolor="cyan">
<form action="verify.jsp" method="post" name="form">
```

```
        请输入管理员账号：
        <input type="text" name="text">
        <p>
        请输入管理员密码：
    <input type="password" name="pwd"/>
    <input type="submit" value="登录">
    </form>
  </body>
  </html>
```

代码模板 verify.jsp 如下：

```
<%@ page language = "java" contentType = "text/html; charset = GBK" pageEncoding =
"GBK"%>
<html>
<head>
    <title>verify.jsp</title>
</head>
<body bgcolor="cyan">
<%
    String strName = request.getParameter("text");
    String strPassword = request.getParameter("pwd");
    if (!("sa".equals(strName) || "admin123".equals(strPassword))) {
        response.sendRedirect("eg4_4.jsp");  //重定向到 eg4_4.jsp 页面重新输入密码
    } else {
        out.print("管理员登录成功,您可以管理页面了!");
    }
%>
</body>
</html>
```

首先运行 eg4_4.jsp 页面，效果如图 4-11 所示。

当用户在该页面中输入正确的管理员账号(即 sa)和正确的管理员密码(即 admin123)，并单击“登录”按钮后，会出现如图 4-12 和图 4-13 所示的页面效果。如果账号和密码输入错误，那么还会重新定向到最初页面。

http://localhost:8080/ch04/eg4_4.jsp

请输入管理员账号：

请输入管理员密码：　登录

图 4-11　例子 4_4 页面效果

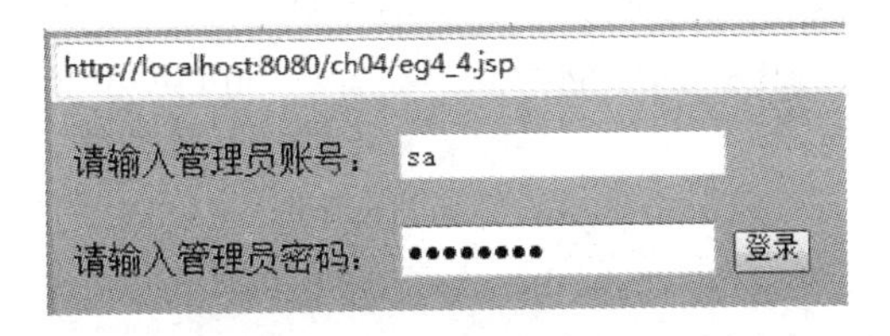

图 4-12　登录页面效果

http://localhost:8080/ch04/verify.jsp

管理员登录成功，您可以管理页面了！

图 4-13　验证页面效果

需要注意的是，在客户端浏览器工作时，Web 服务器要求浏览器重新发送一个请求，用于重定向某个页面，在浏览器地址栏上会出现重定向页面的 URL，且为绝对路径。

在第 3 章学习过的 forward 动作标记也可以实现页面的跳转，例如下面的语句：<jsp:forward page = " verify. jsp"/>。但使用 forward 动作标记与 response 对象调用 sendRedirect 方法有所不同，具体区别分析如下：

(1) forward 为服务器端跳转，浏览器地址栏不变；sendRedirect 为客户端跳转，浏览器地址栏改变为新页面的 URL。

(2) 执行到 forward 标记出现处停止当前 JSP 页面的继续执行，而转向标记中 page 属性指定的页面；sendRedirect 所有代码执行完毕之后再跳转。

(3) 使用 forward，通过 request 请求信息能够保留在下一个页面；使用 sendRedirect 不能保留 request 请求信息。

(4) forward 传递参数的格式如下：

```
<jsp:forward page="verify.jsp">
      <jsp:param name="text" value="sa"/>
      <jsp:param name="pwd" value="admin123"/>
</jsp:forward>
```

response 对象的 sendRedirect 传递参数的方式如下：

```
response.sendRedirect("verify.jsp?text=sa&pwd=admin123");
```

4.2.3 拓展训练

1. 有如下程序段：

```
<form>
    <input type = "text" name = "id">
    <input type= "submit" value ="提交">
</form>
```

下面(　　)语句可以获取用户输入的信息。

A. request. getParameter("id")；　　B. request. getAttribute("submit")；

C. session. getParameter(key, "id")；　　D. session. getAttribute (key, "id")；

2. 下面(　　)内置对象是对客户的请求做出响应，向客户端发送数据的。

A. request　　B. session　　C. response　　D. application

3. 可以使用(　　)方法实现客户的重定向。

A. response. setStatus()；　　B. response. setHeader()；

C. response. setContentType()；　　D. response. sendRedirect()；

4. 什么对象是内置对象？常见的内置对象有哪些？

5. 简述 forward 动作标记与 response. sendRedirect()两种跳转的区别。

6. 修改 4.2.2 节中的例题 eg4_3. jsp,使之具备如下功能：在当前页面停留 5 秒钟后自动跳转到另一个页面 next. jsp(eg4_3. jsp 与 next. jsp 在同一个 Web 服务目录中)。提示：需要为 eg4_3. jsp 设置一个响应头，也就是在 eg4_3. jsp 页面中添加如下代码：

```
response.setHeader("refresh","5;url=next.jsp");
```

7. 编写两个 JSP 页面 input. jsp 和 compute. jsp,用户可以使用 input. jsp 页面提供的表单输入两个运算数，选择一种运算符，并提交给 JSP 页面 compute. jsp 来完成相应的计算功能；如果用户某一项没有输入或者没有选择，都将重新定向到 input. jsp 页面等待用户输入。input. jsp 运行后页面效果如图 4-14 所示，当用户输入运算数、选择运算符后，compute. jsp 页面计算结果如图 4-15 所示。

图 4-14　输入信息

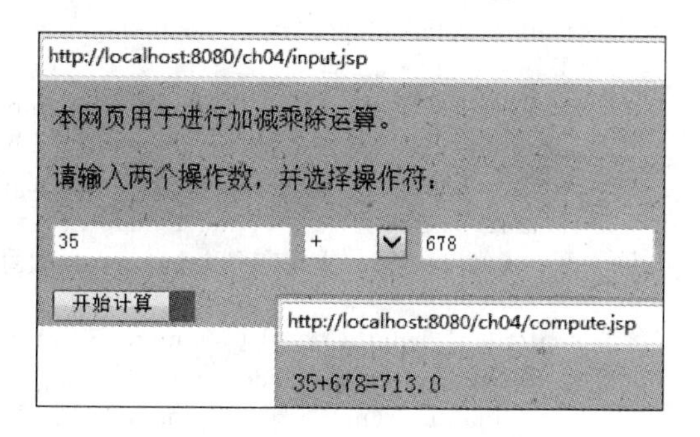

图 4-15　显示计算结果

代码模板 input. jsp 如下：

```
<%@ page language="java" contentType="text/html; charset=GBK" pageEncoding=
"GBK"%>
<html>
<head>
    <title>input.jsp</title>
</head>
<body bgcolor="cyan">
【代码段 1】//定义表单的动作
    本网页用于进行加减乘除运算.
    <p>
    请输入两个操作数,并选择操作符:
    <p>
    <input type="text" name="number1">
    <select size="1" name="list">
        <option value="">操作符
        【代码段 2】//在下拉列表中添加"+"、"-"、"*"、"/"四种运算符
    </select>
```

```
    <input type="text" name="number2">
    <p>
    <input type="submit" value="开始计算" name="button">
  </form>
</body>
</html>
```

代码模板 compute.jsp 如下：

```
<%@ page language = "java" contentType = " text/html; charset = GBK " pageEncoding =
"GBK"%>
<html>
<head>
<title>compute.jsp</title>
</head>
<body bgcolor="cyan">
<%
 String num1 = request.getParameter("number1");
 String num2 = request.getParameter("number2");
 String operator = request.getParameter("list");
 if(num1.length()==0 || num2.length()==0 || operator.length()==0){
      【代码段 3】//重定向到 input.jsp 页面重新输入密码
}
else if(num1.length()!=0 && num2.length()!=0 && operator.length()!=0){
    if(operator.equals("add"))
       out.print(num1+"+"+num2+"="+(Double.parseDouble(num1)+Double.parseDouble
(num2)));
    else if(operator.equals("sub"))
       out.print(num1+"-"+num2+"="+(Double.parseDouble(num1)-Double.parseDouble
(num2)));
    else if(operator.equals("mul"))
       out.print(num1+" * "+num2+"="+(Double.parseDouble(num1) * Double.parseDouble
(num2)));
    else if(operator.equals("div"))
       out.print(num1+"/"+num2+"="+(Double.parseDouble(num1)/Double.parseDouble
(num2)));
    }
%>
</body>
</html>
```

4.3 会话对象 session

客户端浏览器与 Web 服务器之间使用 Http 协议进行通信。Http 是一种无状态协议，客户向服务器发出请求(request)，服务器返回响应(response)，连接就被关闭了，在服务器端不保留连接的相关信息。所以服务器必须采取某种手段来记录每个客户的连接信息。

Web服务器可以使用内置对象session来存放有关连接的信息。session对象指的是客户端与服务器端的一次会话，从客户端连到服务器的一个Web应用程序开始，直到客户端与服务器断开为止。

4.3.1　知识要点

1. session对象的ID

Web服务器会给每一个用户自动创建一个session对象，为每个session对象分配一个唯一标识的String类型的session ID，这个ID用于区分其他用户。这样每个用户都对应着一个session对象(该用户的会话)，不同用户的session对象互不相同。session对象调用getId()方法就可以获取当前session对象的ID。

2. session对象存储数据

使用session对象可以保存用户在访问某个Web服务目录期间的有关数据。处理数据的方法如下：

- public void setAttribute(String key, Object obj)

将参数obj指定的对象保存到session对象中，key为所保存的对象指定一个关键字。若保存的两个对象关键字相同，则先保存的对象被清除。

- public Object getAttibute(String key)

获取session中关键字是key的对象。

- public void removeAttribute(String key)

从session中删除关键字key所对应的对象。

- public Enumeration getAttributeNames()

产生一个枚举对象，该枚举对象可使用方法nextElemets()遍历session中的各个对象所对应的关键字。

3. session对象的生存期限

一个用户在某个Web服务目录中的session对象的生存期限依赖于以下几个因素：

- 用户是否关闭浏览器。
- session对象是否调用invalidate()方法。
- session对象是否达到设置的最长“发呆”时间。

4.3.2　技能操作

具体任务如下：

- 获取session对象的生存期限ID。
- 获取使用session对象存储数据。
- 理解session对象的生存期限。

1．获取 session 对象的 ID

编写三个 JSP 页面 eg4_5.jsp(个人主页)、information.jsp(自我介绍)和 major.jsp(专业介绍)，存放在同一 Web 服务目录中。客户首先访问 eg4_5.jsp 页面，运行结果如图 4-16 所示，从该页面可以链接到 information.jsp 页面，然后从 information.jsp 页面还可以链接到 major.jsp 页面。

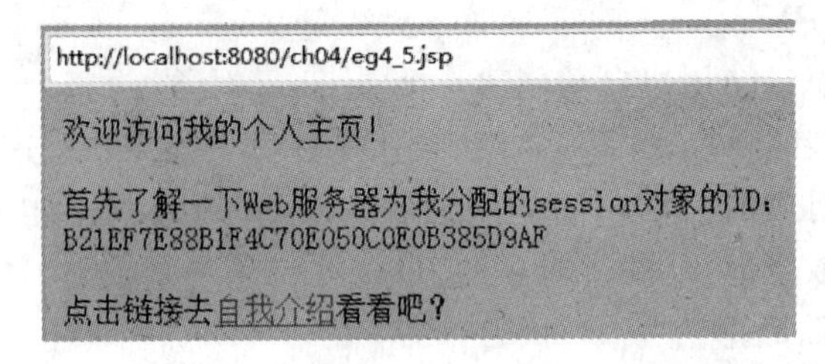

图 4-16　个人主页

代码模板 eg4_5.jsp 如下：

```
<%@ page language="java" contentType="text/html; charset=GBK" pageEncoding=
"GBK"%>
<html>
<head>
<title>eg4_5.jsp</title>
</head>
<body bgcolor="cyan">
    欢迎访问我的个人主页!
    <p>
    首先了解一下 Web 服务器为我分配的 session 对象的 ID:
    <%
        String id= session.getId(); //使用 session 对象调用 getId 方法获得 ID
    %>
    <br>
    <%=id %>
    <p>
    点击链接去<a href="information.jsp">自我介绍</a>看看吧?
</body>
</html>
```

代码模板 information.jsp 如下：

```
<%@ page language="java" contentType="text/html; charset=GBK" pageEncoding=
"GBK"%>
<html>
<head>
<title>information.jsp</title>
</head>
```

```
<body bgcolor="cyan">
    谢谢您关注我的<font size=5>自我介绍</font>!<br><br>
    还是要先来了解一下 Web 服务器为我分配的 session 对象的 ID:
    <%
        String id=session.getId(); //使用 session 对象调用 getId 方法获得 ID
    %>
    <br>
    <%=id %>
    <br><br>
    我生在一个小山村
    那里有我的父老乡亲
    胡子里长满故事
    憨笑中埋着乡音
    <br><br>
    点击链接去<a href="major.jsp">专业介绍</a>看看吧?
</body>
</html>
```

运行结果如图 4-17 所示。

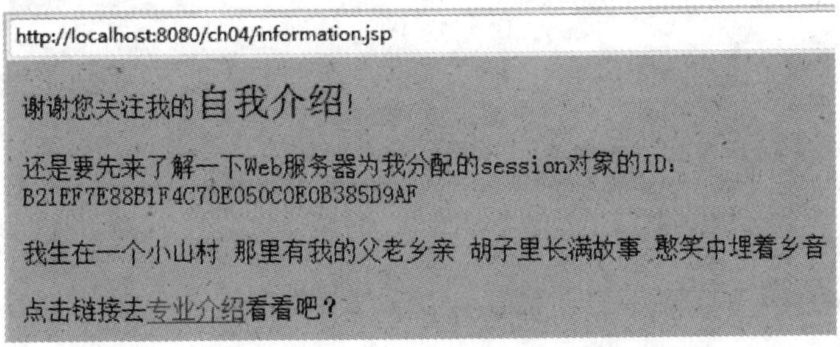

http://localhost:8080/ch04/information.jsp

谢谢您关注我的自我介绍!

还是要先来了解一下Web服务器为我分配的session对象的ID:
B21EF7E88B1F4C70E050C0E0B385D9AF

我生在一个小山村 那里有我的父老乡亲 胡子里长满故事 憨笑中埋着乡音

点击链接去专业介绍看看吧?

图 4-17 自我介绍

代码模板 major.jsp 如下:

```
<%@ page language="java" contentType="text/html; charset=GBK" pageEncoding=
"GBK"%>
<html>
<head>
<title>major.jsp</title>
</head>
<body bgcolor="cyan">
    谢谢您关注我的<font size=5>专业介绍</font>!<br><br>
    最后还是要看看 Web 服务器为我分配的 session 对象的 ID:
    <%
            String id=session.getId(); //使用 session 对象调用 getId 方法获得 ID
    %>
    <br>
```

```
        <%=id %>
        <br><br>
        <h2>计算机专业</h2>
        本专业是计算机硬件与软件相结合、面向系统、侧重应用的宽口径专业。<br>
        通过基础教学与专业训练,培养基础知识扎实、知识面宽、工程实践能力强,<br>
        具有开拓创新意识,在计算机科学与技术领域从事科学研究、教育、开发和应用的高级人才。
        <br>
        本专业开设的主要课程有:程序设计、数据结构、操作系统、计算机网络、数据库系统等。<br>
        <br><br>
        点击链接去<a href="eg4_5.jsp">我的首页</a>看看吧?
    </body>
    </html>
```

运行结果如图 4-18 所示。

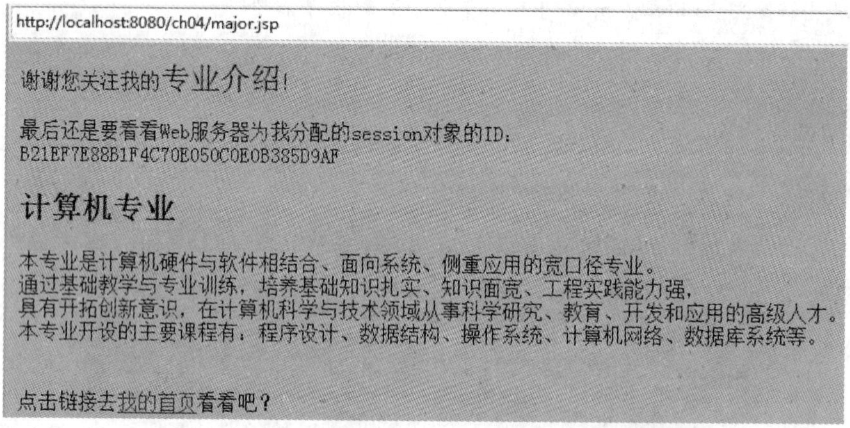

http://localhost:8080/ch04/major.jsp

谢谢您关注我的专业介绍!

最后还是要看看Web服务器为我分配的session对象的ID:
B21EF7E88B1F4C70E050C0E0B385D9AF

计算机专业

本专业是计算机硬件与软件相结合、面向系统、侧重应用的宽口径专业。
通过基础教学与专业训练,培养基础知识扎实、知识面宽、工程实践能力强,
具有开拓创新意识,在计算机科学与技术领域从事科学研究、教育、开发和应用的高级人才。
本专业开设的主要课程有:程序设计、数据结构、操作系统、计算机网络、数据库系统等。

点击链接去我的首页看看吧?

图 4-18 专业介绍

从以上三个页面的运行结果来看,一个用户在同一个 Web 服务目录中只有一个 session 对象,当用户访问相同 Web 服务目录的其他页面时,Web 服务器不会再重新分配 session 对象,直到用户关闭浏览器或这个 session 对象达到了它的生存期限。当用户重新打开浏览器再访问该 Web 服务目录时,Web 服务器为该客户再创建一个新的 session 对象。

需要注意的是,同一用户在多个不同的 Web 服务目录中所对应的 session 对象是不同的,一个服务目录对应一个 session 对象。

2. 使用 session 对象存储数据

编写三个 JSP 页面 eg4_6.jsp、display.jsp 和 feedback.jsp,来模拟学生在线信息填报系统。其中 eg4_6.jsp 页面用于填写个人资料,display.jsp 页面用于显示已填写信息,feedback.jsp 页面用于为已确认提交个人资料的学生显示反馈信息。

代码模板 eg4_6.jsp 如下:

```
<%@ page language = "java" contentType = "text/html; charset = GBK" pageEncoding =
"GBK"%>
<html>
<head>
<title>eg4_6.jsp</title>
</head>
<body bgcolor="cyan">
<form method="post" action="display.jsp">
   <p>学生个人资料填写<br>
   <p>姓名：<input type="text" name="stuName">年龄：<input type="text" name="age">
   <p>籍贯：<input type="text" name="birth">专业：<input type="text" name="major">
   <p>性别：<input type="radio" value="男"ame="sex">男
             <input type="radio" value="女" name="sex">女
       电话：<input type="text" name="phone">
   <p>爱好：<input type="checkbox" name="hobby" value="阅读">阅读
             <input type="checkbox" name="hobby" value="音乐">音乐
             <input type="checkbox" name="hobby" value="电影">电影
             <input type="checkbox" name="hobby" value="上网">上网
             <input type="checkbox" name="hobby" value="游泳">游泳
             <input type="checkbox" name="hobby" value="打球">打球
   <p><input type="submit" value="确定" name="yes">
      <input type="reset" value="重置" name="no">
 </form>
</body>
</html>
```

代码模板 display.jsp 如下：

```
<%@ page language = "java" contentType = "text/html; charset = GBK" pageEncoding =
"GBK"%>
<%@ page import ="java.util.*" %>
<html>
<head>
 <title>display.jsp</title>
</head>
<body bgcolor="cyan">
<%
 request.setCharacterEncoding("GBK");
 String studentName = request.getParameter("stuName");
//下面的语句是将学生姓名 studentName 以"studentName"为关键字存储到 session 对象中
 session.setAttribute("studentName", studentName);
 String stuAge = request.getParameter("age");
 session.setAttribute("stuAge", stuAge);
 String stuBirth = request.getParameter("birth");
 session.setAttribute("stuBirth", stuBirth);
 String stuMajor = request.getParameter("major");
```

```
 session.setAttribute("stuMajor", stuMajor);
 String stuSex = request.getParameter("sex");
 session.setAttribute("stuSex", stuSex);
 String stuPhone = request.getParameter("phone");
 session.setAttribute("stuPhone", stuPhone);
 String stuHobby[]= request.getParameterValues("hobby");
       if(stuHobby!=null)
       { for(int k=0;k<stuHobby.length;k++)
           { session.setAttribute(stuHobby[k],stuHobby[k]);//将相应的关键字存储到session对象中
             }
       }
%>
姓名:<%=studentName%>
年龄:<%=stuAge %><br>
籍贯:<%=stuBirth %>
专业:<%=stuMajor %><br>
性别:<%=stuSex %>
电话:<%=stuPhone %><br>
爱好:
 <%
 for(int i=0;i<stuHobby.length;i++){
       out.print(stuHobby[i]+" ");
 }
 %>
<p>
<form action="feedback.jsp" method="post">
 <input type="submit" value="确认完毕"/>
</form>
<a href="eg4_6.jsp">重新填写</a>
</body>
</html>
```

代码模板 feedback.jsp 如下:

```
<%@ page language = "java" contentType = "text/html; charset = GBK" pageEncoding =
"GBK"%>
<html>
<head>
    <title>feedback.jsp</title>
</head>
<body bgcolor="cyan">
这里是信息反馈页面,
 <%
 //下面的语句是获取session中关键字为studentName的对象
 String name=(String)session.getAttribute("studentName");
 out.print(name);%>
```

```
  的信息已确认提交.
  <p>点击链接去<a href="eg4_6.jsp">该例首页可以修改信息</a>
</body>
</html>
```

运行结果如图4-19所示。

图4-19　信息填写页面效果

当用户填写完信息，并且单击“确定”按钮后，会进入如图4-20所示的信息显示页面。

当用户在信息展示页面中再次单击“确认完毕”按钮后，会出现如图4-21所示的信息反馈页面。

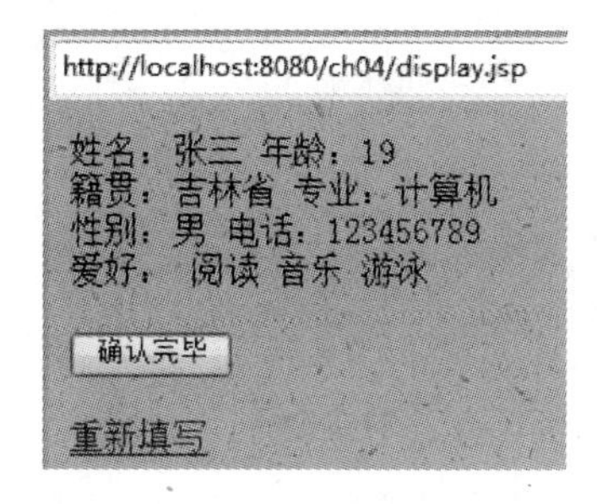

图4-20　信息展示页面

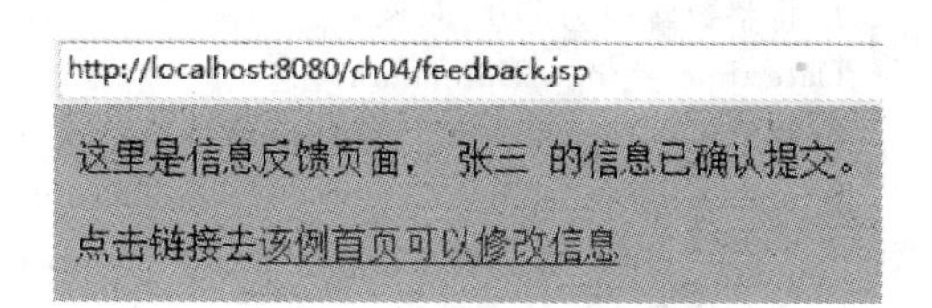

图4-21　信息反馈页面

从上例中可以看出，当某个用户进入某个Web服务目录，服务器就会为该用户分配一个session对象；用户可在该Web服务目录的所有页面使用这个session对象，该session可以存储、获取和移除用户的信息对象；当用户离开该Web服务目录时，session对象随即也会消失。

3. session对象的生存期限

编写一个JSP页面eg4_7.jsp，完成下述功能：用户如果是第一次访问该页面，向用户展示欢迎信息，并输出会话对象允许的最长“发呆”时间、创建时间以及session的ID。在eg4_7.jsp页面中，session对象使用setMaxInactiveInterval(int maxValue)方法设置最长的

“发呆”状态时间为8秒。用户如果两次刷新间隔时间超过8秒,用户先前的session被取消,用户将获得一个新的session对象。

代码模板eg4_7.jsp如下:

```
<%@ page language = " java"  contentType = " text/html; charset = GBK"  pageEncoding =
"GBK"%>
<%@ page import="java.util.*"%>
<%@ page import="java.text.*"%>
<html>
<head>
<meta http-equiv="Content-Type" content="text/html; charset=ISO-8859-1">
    <title>eg4_7.jsp</title>
</head>
<body>
    <%
    //下面语句的功能: session 调用 setMaxInactiveInterval(int n)方法设置最长“发呆”时间为 10 秒
    session.setMaxInactiveInterval(8);
    boolean oper =session.isNew(); //session 调用 isNew()方法判断 session 是不是新创建的
    if (oper) {
        out.println("欢迎您首次访问当前 Web 服务目录 ch04.");
        out.println("<hr/>");
    }
    out.println("session 允许的最长发呆时间为: " +
        session.getMaxInactiveInterval()+ "秒.");
    //获取 session 对象被创建的时间
    long num = session.getCreationTime();
    //将整数转换为 Date 对象
    Date time = new Date(num);
    //用给定的模式和默认语言环境的日期格式符号构造 SimpleDateFormat 对象
    SimpleDateFormat matter = new SimpleDateFormat(
                "北京时间: yyyy 年 MM 月 dd 日 HH 时 mm 分 ss 秒 E.");
    //得到格式化后的字符串
    String strTime = matter.format(time);
    out.println("<br/>session 的创建时间为: " + strTime);
    out.println("<br/>session 的 id 为: " + session.getId() + ".");
 %>
</body>
</html>
```

第一次访问eg4_7.jsp页面,会出现如图4-22所示的页面效果。如果在8秒钟之内刷新该页面,即在8秒钟之内再次访问该页面,会出现如图4-23所示的页面效果;如果超过8秒钟之后再刷新该页面,即在8秒钟之后再次访问该页面,仍会出现类似图4-22所示的效果,但页面中session的id值将与上次出现的id值不同。

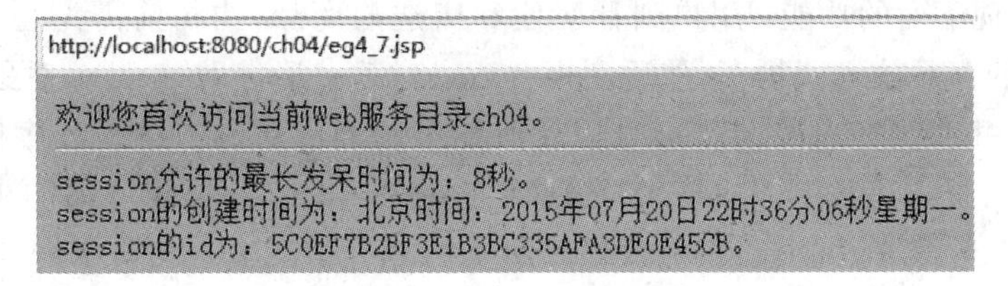

图 4-22　首次访问或间隔 10 秒后访问该页面

http://localhost:8080/ch04/eg4_7.jsp

session允许的最长发呆时间为：8秒。
session的创建时间为：北京时间：2015年07月20日22时36分06秒星期一。
session的id为：5C0EF7B2BF3E1B3BC335AFA3DE0E45CB。

图 4-23　10 秒之内访问该页面

从上例中可以看出,如果用户长时间不关闭浏览器,session 对象也没有调用 invalidate()方法,那么用户的 session 也可能消失。例如 eg4_7.jsp 页面在 8 秒之内不被访问的话,它先前创建的 session 对象就消失了,服务器又重新创建一个 session 对象。这是因为 session 对象达到了它的最大“发呆”时间。所谓“发呆”状态时间,是指用户对某个 Web 服务目录发出的两次请求之间的间隔时间。

用户对某个 Web 服务目录下的 JSP 页面发出请求并得到响应,如果用户不再对该 Web 服务目录发出请求,例如不再操作浏览器,那么用户对该 Web 服务目录进入“发呆”状态,直到用户再次请求该 Web 服务目录时,“发呆”状态结束。

Tomcat 服务器允许用户最长的“发呆”状态时间为 30 分钟。可以通过修改 Tomcat 安装目录中 conf 文件夹下的配置文件 web.xml 文件,找到下面的片断,修改其中的默认值 30,就可以重新设置各个 Web 服务目录下的 session 对象的最长“发呆”时间。这里的时间单位为分钟。

```
<session-config>
        <session-timeout>30</session-timeout>
</session-config>
```

也可以通过 session 对象调用 setMaxInactiveInterval(int time)方法来设置最长“发呆”状态时间,参数的时间单位为秒。

4.3.3 拓展训练

1. 一个用户在不同 Web 服务目录中的 session 对象相同吗？一个用户在同一 Web 服务目录的不同子目录中的 session 对象相同吗？

2. session 对象的生存期限依赖于哪些因素？

3. 编程实现下面的功能：用户到社区便利店选购商品，购物前需先登录会员卡号，购买时先把选购的商品放入购物车，然后到柜台清点付款。请借助 session 对象相关知识来模拟网上购物车，并存储客户的会员卡号以及所购买的商品名称。编写程序模拟上述过程：login.jsp 实现会员卡号输入，lead.jsp 实现商品导购，buy.jsp 实现商品购买，list.jsp 实现商品清点。首先运行 login.jsp，页面效果如图 4-24 所示；当输入会员卡号并单击“登录”按钮后，会员卡号被送出，此时继续在该页单击链接，会进入 lead.jsp 商品导购页面，如图 4-25 所示；在 lead.jsp 商品导购页面中，可以单击“欢迎选购所需要的文具”链接，进入 buy.jsp 商品购物页面，如图 4-26 所示；在 lead.jsp 商品导购页面中，用户可以勾选所购买的商品，然后单击“确定”按钮，进入最终的商品清点页面，如图 4-27 所示。要求将如下代码模板补充完整，并调试运行。

http://localhost:8080/ch04/login.jsp
欢迎光临糖糖社区便利店，请输入您的会员卡号：
780780 登录
单击此处链接可以去购买商品！

图 4-24　会员登录页面

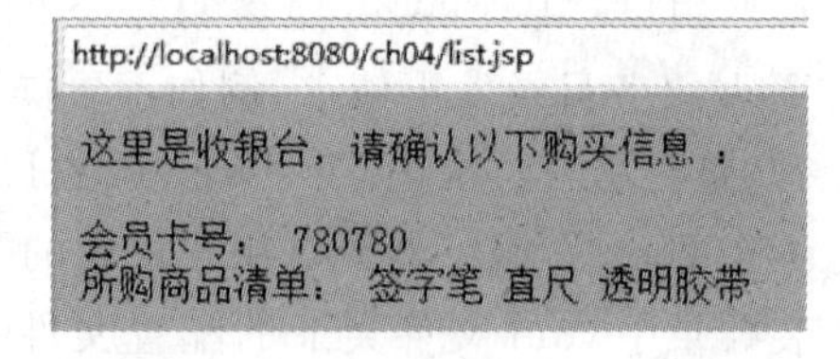

图 4-25　商品导购页面

http://localhost:8080/ch04/buy.jsp
这里是文具柜台，请选择所需文具
☑签字笔 ☑直尺 □卷笔刀 ☑透明胶带 □订书器 □燕尾夹
确定 重置

图 4-26　商品购买页面

http://localhost:8080/ch04/list.jsp
这里是收银台，请确认以下购买信息：
会员卡号：780780
所购商品清单：签字笔 直尺 透明胶带

图 4-27　商品清点页面

代码模板 login.jsp 如下：

```
<%@ page language="java" contentType="text/html; charset=GBK" pageEncoding=
"GBK"%>
<html>
<head>
    <title>login.jsp</title>
</head>
<body bgcolor="cyan">
    <form method="post">
    欢迎光临糖糖社区便利店,请输入您的会员卡号:
    <p>
    <input type="text" name="userID">
    <input type="submit" value="登录" name="button">
```

```
    </form>
    <%
    【代码段 1】//获取用户在文本框 userID 中输入的信息,并存储到字符串对象 userID 中
    if(userID==null)
    { userID=" ";
    }
    else
    {
    【代码段 2】//把会员卡号 userID 以 customerInfo 为关键字存储到 session 对象中
    }
    %>
    <a href="lead.jsp">单击此处链接可以去购买商品!</a>
</body>
</html>
```

代码模板 lead.jsp 如下:

```
<%@ page language="java" contentType="text/html; charset=GBK" pageEncoding=
"GBK"%>
<html>
<head>
    <title>lead.jsp</title>
</head>
<body bgcolor="cyan">
这里是文具柜台:
<p>
<a href="buy.jsp">欢迎选购所需要的文具.</a>
<p>
【代码段 3】//超链接到 login.jsp 页面
您也可以单击这里修改会员卡号信息.</a>
</body>
</html>
```

代码模板 buy.jsp 如下:

```
<%@ page language="java" contentType="text/html; charset=GBK" pageEncoding=
"GBK"%>
<html>
<head>
<title>buy.jsp</title>
</head>
<body bgcolor="cyan">
 <form method="post" action="list.jsp">
        这里是文具柜台,请选择所需文具
        <br>
        <input type="checkbox" name="hobby" value="签字笔">签字笔
```

```
        【代码段 4】//添加"直尺"选项
        <input type="checkbox" name="hobby" value="卷笔刀">卷笔刀
        <input type="checkbox" name="hobby" value="透明胶带">透明胶带
        <input type="checkbox" name="hobby" value="订书器 ">订书器
        <input type="checkbox" name="hobby" value="燕尾夹">燕尾夹
        <p><input type="submit" value="确定" name="yes">
        <input type="reset"value="重置" name="no">
  </form>
</body>
</html>
```

代码模板 list.jsp 如下：

```
<%@ page language = "java" contentType = "text/html; charset = GBK" pageEncoding =
"GBK"%>
<%@ page import ="java.util.*" %>
<html>
<head>
 <title>list.jsp</title>
</head>
<body bgcolor="cyan">
<%
 request.setCharacterEncoding("GBK");
 String foods[]= request.getParameterValues("hobby");
       if(foods!=null)
       { for(int k=0;k<foods.length;k++)
             {【代码段 5】//将 foods[k]相应的关键字存储到 session 对象中
             }
       }
%>
这里是收银台,请确认以下购买信息:
<p>
会员卡号:
<%
【代码段 6】//获取 session 中关键字为 customerInfo 的对象(会员卡号),存入字符串对象 ID 中
out.println(ID);
%>
<br>
所购商品清单:
 <%
 for(int i=0;i<foods.length;i++){
       out.print(foods[i]+" ");
 }
 %>
</body>
</html>
```

4.4　application对象

4.4.1　知识要点

不同用户的session对象互不相同，但有时候用户之间可能需要共享一个对象，Web服务器启动后，就产生了这样一个唯一的内置对象，即application全局应用程序对象。任何用户在访问同一Web服务目录的各个页面时，共享一个application对象，直到服务器关闭，这个application对象被取消为止。application同session对象一样也可以进行数据的存储，处理数据的方法如下：

- public void setAttribute(String key, Object obj)

将参数obj指定的对象保存到application对象中，key为所保存的对象指定一个关键字。若保存的两个对象关键字相同，则先保存的对象被清除。

- public Object getAttribute(String key)

获取application中关键字是key的对象。

- public void removeAttribute(String key)

从application中删除关键字key所对应的对象。

- public Enumeration getAttributeNames()

产生一个枚举对象，该枚举对象可使用方法nextElemets()遍历application中的各个对象所对应的关键字。

4.4.2　技能操作

使用application对象存储数据的功能制作简单的网络聊天室。具体任务描述如下：编写三个JSP页面文件eg4_8_1.jsp、eg4_8_2.jsp和eg4_8_3.jsp。其中eg4_8_1.jsp为聊天室的首页，用户可以在该页输入昵称和聊天内容，同时可将发言内容提交给eg4_8_2.jsp页面；eg4_8_2.jsp页面负责获取用户输入的信息并存储起来；eg4_8_3.jsp负责显示所有用户的聊天内容。

代码模板eg4_8_1.jsp如下：

```
<%@ page contentType="text/html;charset=gb2312" %>
<html>
<body bgcolor="cyan">
  <form action="eg4_8_2.jsp" method="post" name="form">
       欢迎来到简简单单聊天室，畅所欲言！
       <br>昵称：<input type="text" name="peopleName">
       <br>输入您的聊天内容：
       <br><textarea name="messages" rows="10" cols=36 wrap="physical"></textarea>
       <br><input type="submit" value="提交发言" name="submit">
```

```
    </form>
    <form action="eg4_8_3.jsp" method="post" name="form1">
        <input type="submit" value="查看聊天记录" name="look">
    </form>
  </body>
  </html>
```

代码模板 eg4_8_2.jsp 如下：

```
<%@ page contentType="text/html;charset=gb2312" pageEncoding="GBK"%>
<%@ page import="java.util.*" %>
<html><body bgcolor="cyan">
    <%! Vector v=new Vector();
        int i=0; ServletContext application;
        synchronized void sendMessage(String s)
          { application=getServletContext();;
            i++;
          v.add("No."+i+"\t"+s);
//下面语句的功能是把聊天内容 v 以"Mess"为关键字存储到 application 对象中
          application.setAttribute("Mess",v);
         }
    %>
    <% request.setCharacterEncoding("GBK");
    String name=request.getParameter("peopleName");
        String messages=request.getParameter("messages");
            if(name==null)
              {name="guest"+(int)(Math.random()*10000);
              }
            if(messages==null)
              {messages="无信息";
              }
          String s="昵称:"+name+"#"+"聊天内容:"+"<BR>"+messages;
          sendMessage(s);
          out.print("您的发言已经提交!");
    %>
  <a href="eg4_8_1.jsp">返回
</body>
</html>
```

代码模板 eg4_8_3.jsp 如下：

```
<%@ page contentType="text/html;charset=GB2312" %>
<%@ page import="java.util.*" %>
<html><body bgcolor="cyan">
    <%
//下面语句的功能是提取 application 中关键字是 Mess 的对象(聊天内容)
```

```
        Vector v=(Vector)application.getAttribute("Mess");
            for(int i=0;i<v.size();i++)
              { String message=(String)v.elementAt(i);
                StringTokenizer fenxi=new StringTokenizer(message,"#");
                  while(fenxi.hasMoreTokens())
                      { String str=fenxi.nextToken();
                          byte a[]=str.getBytes("GBK");
                          str=new String(a);
                        out.print("<BR>"+str);
                      }
              }
        %>
    </body>
    </html>
```

首先运行 eg4_8_1.jsp,页面效果如图 4-28 所示；当用户输入昵称和聊天内容,单击“提交发言”按钮后,会进入如图 4-29 所示的提示页面；在该提示页面中单击“返回”超链接,会再次回到首页即图 4-28 所示的页面,此时可以单击首页中的“查看聊天内容”按钮,查看聊天内容,如图 4-30 所示。

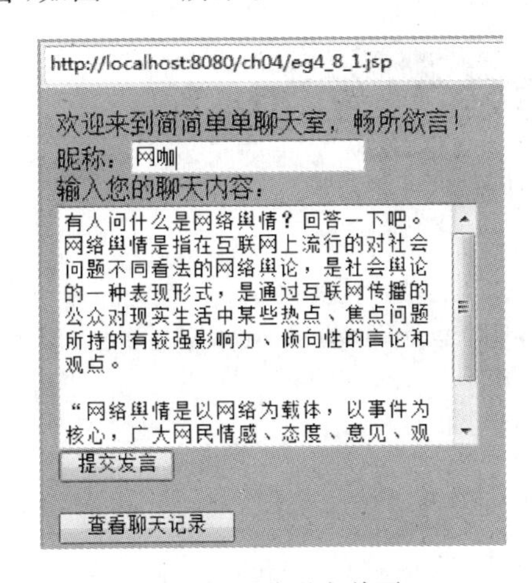

图 4-28　聊天室首页

http://localhost:8080/ch04/eg4_8_2.jsp

您的发言已经提交！　返回

图 4-29　提示页面

http://localhost:8080/ch04/eg4_8_3.jsp

No.1 昵称:网咖
聊天内容:
有人问什么是网络舆情？回答一下吧。网络舆情是指在互联网上流行的对社会问题不同看法的网络舆论,
众对现实生活中某些热点、焦点问题所持的有较强影响力、倾向性的言论和观点。　“网络舆情是以网络
观点的表达、传播与互动，以及后续影响力的集合。”

图 4-30　查看聊天记录页面

上例 eg4_8_2.jsp 文件中的 sendMessage 方法之所以定义为同步方法，是因为 application 对象针对所有的用户都是同等的，任何用户对该对象中存储的数据的操作都会影响到其他用户。如果客户浏览不同的 Web 服务目录，将产生不同的 application 对象。同一个 Web 服务目录中的所有 JSP 页面都共享同一个 application 对象，即使浏览这些 JSP 页面的是不同的用户也是如此。因此，保存在 application 对象中的数据不仅可以跨页面分享，还可以由所有用户共享。

有些 Web 服务器不能直接使用 application 对象，必须使用父类 ServletContext 声明这个对象，然后使用 getServletContext()方法为 application 对象进行实例化。例如该任务中 eg4_8_2.jsp 页面中的代码。

4.4.3 拓展训练

1. 请简述内置对象 request、session 和 application 之间的区别。

2. 模仿 4.4.2 节中的例题，使用 application 对象实现网站访客计数器的功能。访客计数器运行页面效果如图 4-31 所示。要求将代码模板补充完整，并调试运行。

http://localhost:8080/ch04/

您好！欢迎来访，你是第1位到访者

图 4-31　访客计数器

代码模板如下：

```
<%@ page contentType="text/html;charset=gb2312" pageEncoding="GBK"%>
<html>
<body bgcolor="cyan">
<%
【代码段 1】//提取 application 中关键字是 count 的对象(计数内容)
if(count==null){
    count="1";
}else{
    count=Integer.parseInt(count)+1+"";
}
【代码段 2】//把计数内容 count 以关键字"count"为关键字存储到 application 对象中
%>
<%="<h1>您好!欢迎来访,你是第"+count+"位到访者</h1><br>"%>
</body>
 </html>
```

4.5 小　　结

- 所有的内置对象不需要由JSP的编写者声明和实例化，可以直接在所有的JSP网页中使用。内置对象只在Java程序片或者Java表达式中使用。
- 请求对象request代表客户端发出的请求信息对象，通常用来获取表单上的信息。request对象只在从客户发出请求到服务器做出响应这个期间是有效的。
- 应答对象response代表从服务器端返回给客户端的响应信息对象。response对象包含从动态页面返回给客户的所有信息。response可以用来进行页面的重定向，但与用forward动作标记实现页面的跳转有所不同。
- 会话对象session可以用来保存每个用户信息，以便跟踪每个用户的操作状态，不同用户的session对象互不相同。session对象主要用来存储和获取数据，使用时要注意其生存期限。
- 全局应用程序对象application显示相应网页所有应用程序的对象。任何用户在访问同一Web服务目录的各个页面时，共享一个application对象，直到服务器关闭，这个application对象被取消为止。保存在application对象中的数据不仅可以跨页面分享，还可以由所有用户共享。

第 5 章

JavaBean 的使用

通过前几章的学习，我们已经知道一个 JSP 页面通过使用 HTML 标记为用户显示数据(静态部分)，而页面中变量的声明、程序片以及表达式为用户处理数据(动态部分)。如果 Java 程序片和 HTML 标记大量掺杂在一起使用，就不利于 JSP 页面的扩展和维护。JSP 和 JavaBean 技术的结合不仅可以实现数据的表示和处理分离，而且可以提高 JSP 程序代码重用的程度，是 JSP 编程中常用的技术。

在本章中，我们新建一个 Web 工程 ch05，本章例子中涉及的 Java 源文件保存在 ch05 的 src 中，涉及的 JSP 页面保存在 ch05 的 WebContent 中。

5.1 编写 JavaBean

5.1.1 知识要点

JavaBean 是一个可重复使用的软件组件，是遵循一定标准、用 Java 语言编写的一个类，该类的一个实例称为一个 JavaBean，简称 bean。

由于 JavaBean 是基于 Java 语言的，因此 JavaBean 具有以下特点：

- 与平台无关。
- 代码的重复利用。
- 易扩展、易维护、易使用。

编写一个 JavaBean 就是编写一个 Java 的类(该类必须带有包名)，这个类创建的一个对象称为一个 bean，为了让 JSP 引擎(例如 Tomcat)知道这个 bean 的属性和方法，必须在类的方法命名上遵守以下规则：

(1) 如果类的成员变量的名字是 name，那么为了获取或更改成员变量的值，类中必须提供两个方法：

- getName()——用来获取属性 name。
- setName()——用来设置属性 name。

即方法的名字用 get 或 set 为前缀，后缀是首字母大写的成员变量的名字。

(2) 对于 boolean 类型的成员变量，允许使用“is”代替上面的“get”和“set”。

(3) 类中声明的方法的访问权限都必须是 public。

(4) 类中声明的构造方法必须是 public、无参数。

5.1.2　技能操作

使用JavaBean的编写规则编写创建bean的Java源文件。具体任务如下：

- 在Eclipse中创建一个类Student，属于china.dalian包。
- 编写bean的源文件Student.java（在包china.dalian中），该bean的作用是求学生数学课和外语课的总分。

1. 在Eclipse中创建Student类

在Eclipse中，右键单击ch05项目，然后在弹出的快捷菜单中选择New->Class命令，会弹出新建类对话框，在该对话框中可以设置包名和类名，如图5-1所示。设置完以后的目录结构如图5-2所示。

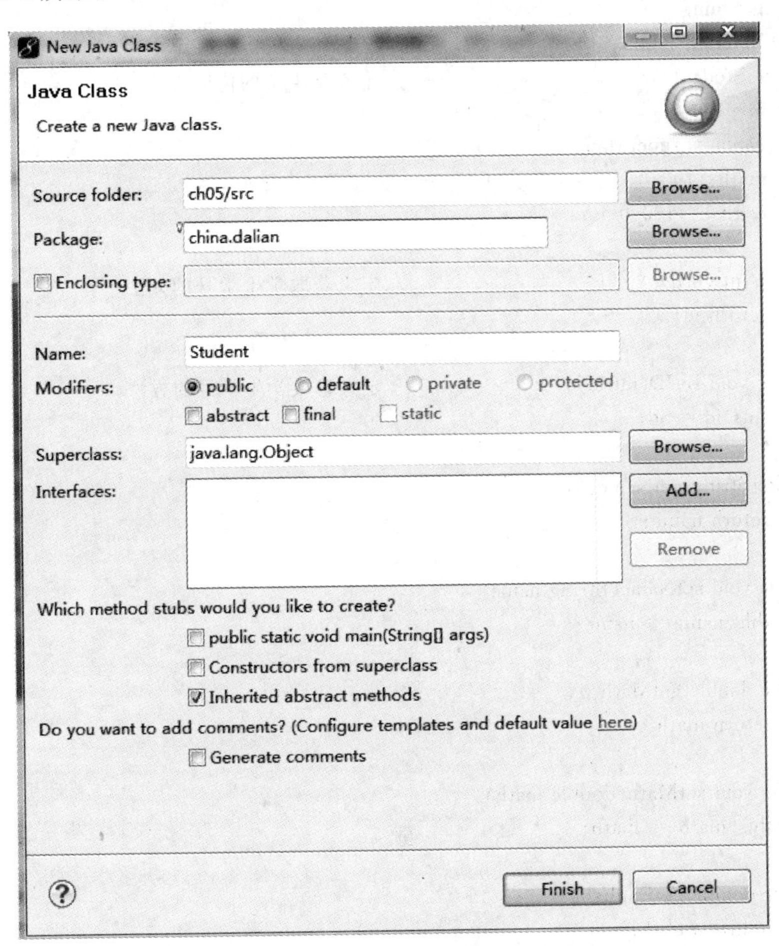

图5-1　新建类对话框

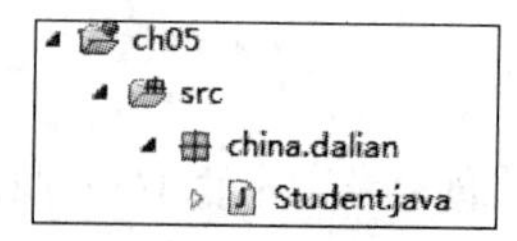

图 5-2 目录结构

2. 编写 bean 的源文件 Student. java

代码模板 Student. java 如下：

```
package china.dalian;
public class Student{
    int id;
    String name;
    double math,english;
    public Student(){                          //定义学生类的构造方法
        id=1;
        name="superMan";
        math=100.0;
        english=100.0;
    }
    public int getID(){                        //定义获取学生学号的方法
        return id;
    }
    public void setID(int id){                 //定义修改学生学号的方法
        this.id = id;
    }
    public String getName(){
        return name;
    }
    public void setName(String name){
        this.name = name;
    }
    public double getMath(){
        return math;
    }
    public void setMath(double math){
        this.math = math;
    }
    public double getEnglish(){
        return english;
    }
    public void setEnglish(double english){
        this.english = english;
```

```
    }
    public double getSum(){
        return math+english;
    }
}
```

JavaBean可以在任何Java程序编辑环境下完成编写,再通过编译生成一个字节码文件(.class文件),为了让JSP引擎(例如Tomcat)找到这个字节码,必须把字节码文件放在特定的位置。本书使用Eclipse集成环境开发JSP程序,Java类的字节码文件由Eclipse自动保存到Web工程的build\classes中。例如,该任务中的Student.class文件保存在ch05\build\classes\china\dalian目录中。

5.1.3 拓展训练

1. JavaBean中声明的方法的访问属性必须是(　　)。

   A. private　　B. public　　C. protected　　D. friendly

2. 写一个bean时,与布尔逻辑类型的成员变量xxx对应的方法是(　　)。

   A. getXxx()　　B. setXxx()　　C. Xxx()　　D. isXxx()

3. 创建bean的源文件Rect.java(在包graph.picture中),该bean的作用是计算矩形的周长和面积。要求将下面的代码模板补充完整,并进行调试。

代码模板Rect.java如下:

```
package graph.picture;
public class Rect {
 double length;
 double width;
【代码段1】{                    //定义类Rect的构造方法
    length=20.56;
    width=15.45;
 }
【代码段2】{                    //定义获取矩形长度的方法
    return length;
 }
【代码段3】{                    //定义修改矩形长度的方法
    this.length = length;
 }
 public double getWidth() {
    return width;
 }
 public void setWidth(double width) {
    this.width = width;
 }
```

```
    public double getGirth(){
        return (length+width) * 2;
    }
    public double getArea(){
       【代码段 4】//返回矩形的面积
    }
}
```

5.2 使用 JavaBean

5.2.1 知识要点

在 JSP 页面中使用 bean，首先必须使用 page 指令的 import 属性导入创建 bean 的类所在的包，例如：

```
<%@ page import=" china.dalian. * "%>
```

然后使用 JSP 动作标记 useBean，来创建与使用 bean。useBean 标记的格式为：

```
<jsp:useBean id= "bean 的名字" class= "创建 bean 的类" scope= "bean 的有效范围"/>
```

或

```
<jsp:useBean id= "bean 的名字" class= "创建 bean 的类" scope= "bean 的有效范围"></jsp:useBean>
```

例如：

```
<jsp:useBean id= "rectangle" class= "small.dog.Rectangle" scope= "page"/>
```

尤其需要说明的是，useBean 标记中 scope 的默认值是 page，除 page 之外，scope 的取值还有 request、session 与 application。

1. scope 取值 page

该 bean 的有效范围是当前页面。当客户请求 bean 时，分配内存空间给它，当客户离开这个页面时，便取消分配的 bean，并收回内存空间。JSP 引擎分配给每个 JSP 页面的 bean 是不同的，它们占有不同的内存空间。

当两个客户访问同一个 JSP 页面时，一个用户对自己 bean 的属性的改变，不会影响到另一个客户。

2. scope 取值 request

该 bean 的有效范围是 request 期间。客户在网站访问时请求多个页面，如果每个页面都含有 useBean 标记，那么在每个页面分配的 bean 也不相同。JSP 引擎对请求做出响应后，bean 将消失。

当两个客户同时请求一个 JSP 页面时，一个用户对自己 bean 属性的改变，不会影响另外一个客户。

3. scope 取值 session

该 bean 的有效范围是客户的会话期间。如果客户在多个页面中互相连接，每个页面都含有一个相同的 useBean 标记，那么这个客户在这些页面得到的 bean 是相同的，即占有相同的内存空间。当会话结束时，bean 消失，释放空间。

如果一个客户在某个页面更改了 bean 的某个属性，那么该客户的其他页面 bean 的属性也发生变化。但两个客户同时访问一个 JSP 页面时，一个客户对自己 bean 的属性的改变不会影响到另一个客户。

4. scope 取值 application

该 bean 的有效范围是 application 期间（Web 服务器启动期间）。JSP 引擎为所有的 JSP 页面分配一个共享的 bean。

当几个客户同时访问一个 JSP 页面时，任何一个客户对自己 bean 的属性的改变都会影响到其他客户。

当含有 useBean 动作标记的 JSP 页面被 JSP 引擎（例如 Tomcat）加载执行时，JSP 引擎首先根据 id 的名字，在 pageContext 内置对象中查看是否含有名字 id 和作用域 scope 的对象；如果该对象存在，JSP 引擎就将这个对象的副本（bean）分配给 JSP 页面使用；如果没有找到，就根据 class 指定的类创建一个名字是 id 的 bean，并添加到 pageContext 对象中，同时将这个 bean 分配给 JSP 页面使用。useBean 动作标记执行流程如图 5-3 所示。

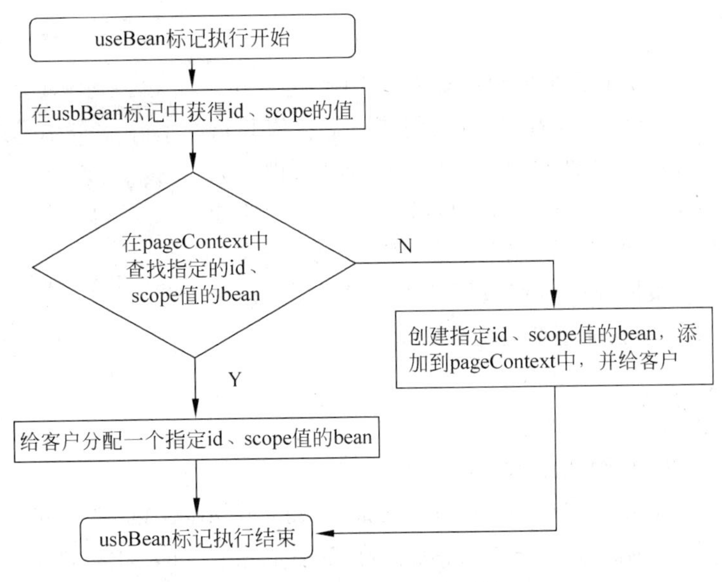

图 5-3　useBean 标记执行流程图

5.2.2 技能操作

在 JSP 页面中使用动作标记 useBean，具体任务如下：编写一个 JSP 页面 eg5_1.jsp，在 JSP 页面中使用 useBean 标记获得一个 bean，负责创建 bean 的类是 5.1.2 节任务中的 Student 类，创建 bean 的名字是 student，student 的 scope 取值为 page。

代码模板 eg5_1.jsp 如下：

```
<%@ page language = "java" contentType = "text/html; charset = GBK" pageEncoding =
"GBK"%>
<%@ page import="china.dalian.Student"%>
<html>
<head>
 <title>eg5_1.jsp</title>
</head>
<body bgcolor="cyan">
<%--通过 useBean 标记获得一个 bean,负责创建 bean 的类是 china.dalian.Student,id 是 student,
scope 取值为 page --%>
 <jsp:useBean id="student" class="china.dalian.Student" scope="page"/>
 <p>该生的学号是: <%=student.getID()%>
 <p>姓名是: <%=student.getName() %>
 <p>数学成绩是: <%=student.getMath() %>
 <p>英语成绩是: <%=student.getEnglish() %>
 <p>总成绩是: <%=student.getSum() %>
</body>
</html>
```

运行上面的 eg5_1.jsp，页面效果如图 5-4 所示。

从创建 bean 的过程可以看出，首次创建一个新的 bean 需要用相应的字节码文件创建对象，当别的 JSP 页面再需要同样的 bean 时，JSP 引擎直接将 pageContext 内置对象里已经存在的对象的副本分配给相应的 JSP 页面，提高了代码的复用程度。如果程序员修改了字节码文件，必须重启 JSP 引擎，才能使用新的字节码文件。

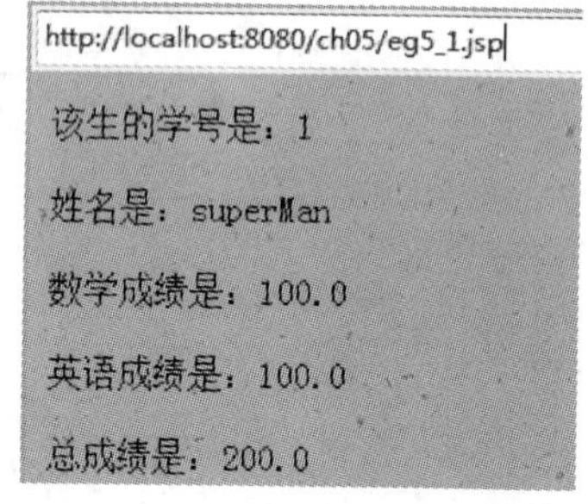

图 5-4 使用 bean 的 JSP 页面

5.2.3 拓展训练

1. 下面(　　)是正确使用 JavaBean 的方式。

   A. <jsp:useBean id="address" class="tom.AddressBean" scope="page"/>

   B. <jsp:useBean name="address" class="tom.AddressBean" scope="page"/>

   C. <jsp:useBean bean="address" class="tom.AddressBean" scope="page" />

   D. <jsp:useBean beanName="address" class="AddressBean" scope="page" />

2. JavaBean的作用域可以是(　　)、page、session和application。

A. request　　B. response　　C. out　　D. 以上都不对

3. 在J2EE,test.jsp文件中有如下一行代码:

```
<jsp:useBean class="tom.jiafei.Test" id="user" scope="________" />
```

要使user对象一直存在于会话中,直至终止或被删除为止,下划线中应填入(　　)。

A. page　　B. request　　C. session　　D. application

4. 关于JavaBean正确的说法是(　　)。

A. 类中声明的方法的访问权限必须是public

B. 在JSP文件中引用bean,其实就是用<jsp:useBean>语句

C. bean文件放在任何目录下都可以被引用

D. 以上均不对

5. 在J2EE中,test.jsp文件中有如下一行代码:

```
<jsp:useBean id=" user" scope="________" type=" com.UserBean" />
```

要使user对象在用户对其发出请求时存在,下划线中应填入(　　)。

A. page　　B. request　　C. session　　D. application

6. 在JSP中,使用<jsp:useBean>动作可以将JavaBean引入JSP页面,对JavaBean的访问范围不能是(　　)。

A. page　　B. request　　C. response　　D. application

7. 在J2EE中test.jsp文件如下:

```
<body>
<jsp:useBean id ="buffer" scope="page" class="java.lang.StringBuffer"/>
<% buffer.append("ABC"); %>
  buffer is <%=buffer%>
</body>
```

该文件试图运行时,将发生(　　)。

A. 编译期间发生错误

B. 运行期间抛出异常

C. 运行后,浏览器上显示: buffer is null

D. 运行后,浏览器上显示: buffer is ABC

8. 编写一个JSP页面computerRect.jsp,在JSP页面中使用useBean标记获得一个bean,负责创建bean的类是5.1.3节中的Rect类,创建bean的名字为rect,rect的scope取值为request。JSP页面的运行效果如图5-5所示。要求将如下代码模板补充完整,并调试运行。

图5-5　使用bean计算矩形的周长和面积

代码模板 test.jsp 如下：

```
<%@ page language = "java" contentType = "text/html; charset = GBK" pageEncoding =
"GBK"%>
【代码段 1】//使用 page 指令标记的 import 属性,导入 Rect 类
<head>
<title>test.jsp</title>
</head>
<body bgcolor="cyan">
【代码段 2】//<%--通过 useBean 标记获得一个 bean,负责创建 bean 的类是 graph.picture.Rect,id
是 rect,scope 取值为 page --%>
 <p>矩形的长是: <%=rect.getLength()%>
 <p>矩形的宽是: <%=rect.getWidth() %>
 <p>矩形的周长是: <%=rect.getGirth() %>
 <p>矩形的面积是: <%=rect.getArea() %>
</body>
</html>
```

5.3 获取 bean 属性

JavaBean 的实质是遵守一定规范的类所创建的对象,可以通过以下两种方式获取 bean 的属性：

1. Java 程序片

可以通过调用构造方法获得一个 bean,然后调用 getXxx()方法来获取 bean 的属性。

2. JSP 标记

先通过<jsp:useBean>标记获得一个 bean,再通过<jsp:getProperty>标记获取 bean 的属性值(无须使用 Java 程序片)。

5.3.1 知识要点

使用 getProperty 动作标记可以获得 bean 的属性值。使用该动作标记之前,必须事先使用 useBean 动作标记获得一个相应的 bean。getProperty 动作标记语法格式如下：

```
<jsp:getProperty name= "bean 的名字" property= "bean 的属性" />
```

或

```
<jsp:getProperty name= "bean 的名字" property= "bean 的属性"/></jsp:getProperty>
```

其中,name 取值是 bean 的名字,和 useBean 标记中的 id 对应; property 取值是 bean 的一个属性的名字,和创建该 bean 的类的成员变量名对应。这条指令相当于在 Java 表达式中使用 bean 的名字调用 getXxx 方法。

5.3.2 技能操作

使用getProperty动作标记获得bean的属性，具体任务如下：

- 创建bean的源文件NewStudent.java(在包china.dalian中)，该bean的作用是计算学生三门课程(数学、英语、政治)的成绩总分和平均分。
- 编写一个JSP页面useGetProperty.jsp，在该JSP页面中使用useBean标记创建一个名字是score的bean，并使用getProperty动作标记获取score的每个属性的值。负责创建score的类是NewStudent类。

1. 创建bean的源文件NewStudent.java

代码模板NewStudent.java如下：

```
package china.dalian;
public class NewStudent{
    int number;
    String name;
    double math,english,politics;
    double sum=0.0,average=0.0;
    public NewStudent(){                              //定义学生类的构造方法
        number=1;
        name="superMan";
        math=100.0;
        english=100.0;
        politics=100.0;
    }
    public int getNumber(){                           //定义获取学生学号的方法
        return number;
    }
    public void setNumber(int number){                //定义修改学生学号的方法
        this.number = number;
    }
    public String getName(){
        return name;
    }
    public void setName(String name){
        this.name = name;
    }
    public double getMath(){
        return math;
    }
    public void setMath(double math){
        this.math = math;
    }
    public double getEnglish(){
```

```
            return english;
        }
        public void setEnglish(double english){
            this.english = english;
        }
        public double getPolitics(){
            return politics;
        }
        public void setPolitics(double politics){
            this.politics = politics;
        }
        public double getSum(){
            sum=math+english+politics;
            return sum;
        }
        public double getAverage(){
            average=sum/3;
            return average;
        }
}
```

2. 编写 JSP 页面 useGetProperty.jsp

代码模板 useGetProperty.jsp 如下：

```
<%@ page language = "java" contentType = "text/html; charset = GBK" pageEncoding =
"GBK"%>
<%@ page import="china.dalian.NewStudent"%>
<html>
<head>
<title>useGetProperty.jsp</title>
</head>
<body bgcolor="cyan">
<h2>这是使用 getProperty 标记的页面。</h2>
<%--下面的语句功能是通过 useBean 标记获得一个 bean,负责创建 bean 的类是 china.dalian
.NewStudent,id 是 score,scope 取值为 page --%>
    <jsp:useBean id="score" class="china.dalian.NewStudent" scope="page"/>
    <%score.setNumber(10);%>
    <%score.setName("littlePerson");%>
    <p>该生的学号是: <jsp:getProperty property="number" name="score"/>
        姓名是: <jsp:getProperty property="name" name="score"/>
    <p>数学成绩是: <jsp:getProperty property="math" name="score"/>
        英语成绩是: <jsp:getProperty property="english" name="score"/>
        政治成绩是: <jsp:getProperty property="politics" name="score"/>
    <p>总成绩是: <jsp:getProperty property="sum" name="score"/>
        平均分是: <jsp:getProperty property="average" name="score"/>
</body>
</html>
```

运行 JSP 页面 useGetProperty. jsp,效果如图 5-6 所示。

http://localhost:8080/ch05/useGetProperty.jsp

这是使用getProperty标记的页面。

该生的学号是：10 姓名是：littlePerson

数学成绩是：100.0 英语成绩是：100.0 政治成绩是：100.0

总成绩是：300.0 平均分是：100.0

图 5-6　使用 getProperty 标记获得 bean 的属性值

从上例中可以看出,在 JSP 页面中使用 getProperty 动作标记获得 bean 的属性时,必须保证 bean 中有相应的 getXxx 方法,即创建 bean 的类中定义 getXxx 方法。

从 useGetProperty. jsp 页面可以看出,使用 getProperty 动作标记获得 bean 的属性值,减少了 Java 程序片的使用。

在 NewStudent. java 中,有两个属性(sum 和 average)没有提供 set 方法,因为二者依赖于 math、english 和 politics 属性。

5.3.3　拓展训练

1. 下面的语句中,与<jsp:getProperty name="aBean" property="jsp"/>等价的是(　　)。

   A. <%=jsp()%>

   B. <%out. print(aBean. getJsp());%>

   C. <%=aBean. setJsp()%>

   D. <%aBean. setJsp();%>

2. 在 JSP 中使用<jsp:getProperty>标记时,不会出现的属性是(　　)。

   A. name　　　　B. property

   C. value　　　　D. 以上皆不会出现

3. 创建 bean 的源文件 Ladder. java(在包 graph. picture 中),该 bean 的作用是计算梯形的面积；编写一个 JSP 页面 ladderProperty. jsp,在该 JSP 页面中使用 useBean 标记创建一个名字是 lad 的 bean,并使用 getProperty 动作标记获得 lad 的每个属性的值。负责创建 lad 的类是 Ladder 类。

4. 创建 bean 的源文件 Circle. java(在包 graph. picture 中),该 bean 的作用是计算圆形的周长和面积；编写一个 JSP 页面 circleProperty. jsp,在该 JSP 页面中使用 useBean 标记创建一个名字是 circle 的 bean,并使用 getProperty 动作标记获得 circle 的每个属性的值。负责创建 circle 的类是 Circle 类。

5.4 设置 bean 属性

我们已经知道，可以通过两种方式获取 bean 的属性，同样修改 bean 的属性也有以下两种方式：

1. Java 程序片

可以通过调用构造方法获得一个 bean，然后调用 setXxx()方法来修改 bean 的属性。

2. JSP 标记

先通过<jsp:useBean>标记获得一个 bean，再通过<jsp:setProperty>标记修改 bean 属性值。

5.4.1 知识要点

使用 setProperty 动作标记可以修改 bean 的属性值。使用该动作标记之前，必须事先使用 useBean 动作标记获得一个相应的 bean。使用 setProperty 动作标记进行 bean 属性的设置有三种方式：

1. 用表达式或字符串设置 bean 的属性

(1) 用表达式设置 bean 的属性。

```
<jsp:setProperty name="bean的名字" property= "bean的属性" value= "<%=expression%>" />
```

(2) 用字符串设置 bean 的属性。

```
<jsp:setProperty name="bean的名字" property= "bean的属性" value=字符串/>
```

2. 通过 HTTP 表单的参数的值设置 bean 的属性

```
<jsp:setProperty name="bean的名字" property=" * " />
```

3. 任意指定请求参数设置 bean 的属性

```
<jsp:setProperty name="bean的名字" property="属性名" param="参数名"/>
```

可以根据自己的需要，任意选择传递的参数，请求参数名无须与 bean 属性名相同。

5.4.2 技能操作

使用 setProperty 动作标记修改 bean 的属性，具体任务如下：

- 用表达式或字符串修改 bean 的属性。
- 通过 HTTP 表单的参数的值设置 bean 的属性。

1. 用表达式或字符串修改 bean 的属性

使用 5.3.2 节中创建的 bean 源文件 NewStudent.java(在包 china.dalian 包中)，前面

已经介绍过，该 bean 的作用是计算学生三门课程(数学、英语、政治)的成绩总分和平均分。然后创建 JSP 页面 useSetProperty_1.jsp，在该 JSP 页面中使用 useBean 标记创建一个名字是 score 的 bean，其有效范围是 page，并使用动作标记修改、获取该 bean 的属性的值。负责创建 score 的类是 NewStudent 类。其中 useSetProperty_1.jsp 的页面效果如图 5-7 所示。

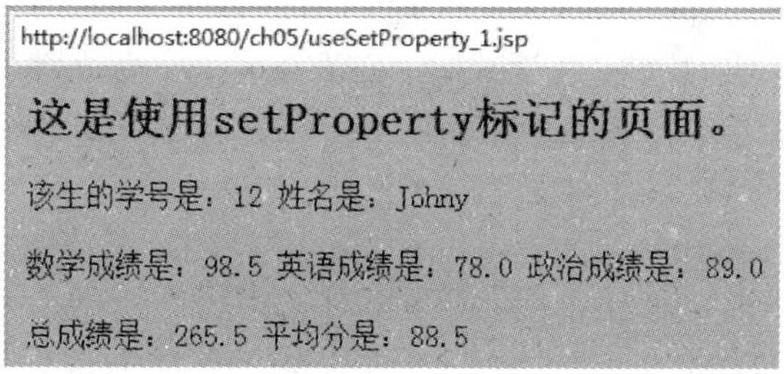

http://localhost:8080/ch05/useSetProperty_1.jsp

这是使用setProperty标记的页面。

该生的学号是：12 姓名是：Johny

数学成绩是：98.5 英语成绩是：78.0 政治成绩是：89.0

总成绩是：265.5 平均分是：88.5

图 5-7　使用字符串或表达式的值修改 bean 的属性

代码模板 NewStudent.java：

参见 5.3.2 节中的 NewStudent.java。

代码模板 useSetProperty_1.jsp 如下：

```
<%@ page language = "java" contentType = " text/html; charset = GBK" pageEncoding =
"GBK"%>
<%@ page import="china.dalian.NewStudent"%>
<html>
<head>
<title>useSetProperty_1.jsp</title>
</head>
<body bgcolor="cyan">
<h2>这是使用 setProperty 标记的页面.</h2>
<jsp:useBean id="score" class="china.dalian.NewStudent" scope="page"/>
   <%--使用 setProperty 标记设置 score 的各个属性值--%>
   <jsp:setProperty property="number" name="score" value="12"/>
   <jsp:setProperty property="name" name="score" value="Johny"/>
   <jsp:setProperty property="math" name="score" value="98.5"/>
   <jsp:setProperty property="english" name="score" value="78.0"/>
   <jsp:setProperty property="politics" name="score" value="89.0"/>
   <p>该生的学号是: <jsp:getProperty property="number" name="score"/>
         姓名是: <jsp:getProperty property="name" name="score"/>
   <p>数学成绩是: <jsp:getProperty property="math" name="score"/>
         英语成绩是: <jsp:getProperty property="english" name="score"/>
         政治成绩是: <jsp:getProperty property="politics" name="score"/>
   <p>总成绩是: <jsp:getProperty property="sum" name="score"/>
         平均分是: <jsp:getProperty property="average" name="score"/>
</body>
</html>
```

用表达式修改 bean 属性的值时，表达式值的类型必须与 bean 的属性类型一致。用字符串修改 bean 属性的值时，字符串会自动被转化为 bean 的属性的类型，不能转化成功的可能会抛出 NumberFormatException 异常。

2. 通过 HTTP 表单的参数的值设置 bean 的属性

编写两个 JSP 页面：inputMess.jsp 和 showMess.jsp。在 inputMess.jsp 页面中可以输入学生信息（包括学号、姓名、数学成绩、英语成绩、政治成绩），输入完毕后提交给 showMess.jsp 页面显示信息。页面中用到的 bean 仍然是使用 5.3.2 节中的 NewStudent 类来创建

代码模板 NewStudent.java：

参见 5.3.2 节中的 NewStudent.java。

代码模板 inputMess.jsp

```
<%@ page language = "java" contentType = "text/html; charset = GBK" pageEncoding =
"GBK"%>
<html>
<head>
<title>inputMess.jsp</title>
</head>
<body bgcolor="cyan">
 <form action="showMess.jsp" method="post">
       请输入学生学号：<input type="text" name="number"/><br>
       请输入学生姓名：<input type="text" name="name"/><br>
       请输入该生数学课程分数：<input type="text" name="math"/><br>
       请输入该生英语课程分数：<input type="text" name="english"/><br>
       请输入该生政治课程分数：<input type="text" name="politics"/><br>
       <input type="submit" value="提交"/>
       <input type="reset" value="重置"/>
 </form>
</body>
</html>
```

代码模板 showMess.jsp 如下：

```
<%@ page language = "java" contentType = "text/html; charset = GBK" pageEncoding =
"GBK"%>
<%@ page import="china.dalian.NewStudent"%>
<%
request.setCharacterEncoding("GBK");
%>
<html>
<head>
<title>showMess.jsp</title>
</head>
```

```
<body>
    <jsp:useBean id="student" class="china.dalian.NewStudent" scope="page"/>
    <%--通过 HTTP 表单的参数的值设置 bean 的属性(表单参数与属性自动匹配)--%>
    <jsp:setProperty property=" * " name="student"/>
    <p>该生的学号是: <jsp:getProperty property="number" name="student"/>
        姓名是: <jsp:getProperty property="name" name="student"/>
    <p>数学成绩是: <jsp:getProperty property="math" name="student"/>
        英语成绩是: <jsp:getProperty property="english" name="student"/>
        政治成绩是: <jsp:getProperty property="politics" name="student"/>
    <p>总成绩是: <jsp:getProperty property="sum" name="student"/>
        平均分是: <jsp:getProperty property="average" name="student"/>
</body>
</html>
```

首先运行 inputMess.jsp 页面，效果如图 5-8 所示；当用户输入完信息单击“提交”按钮后，会进入 showMess.jsp 页面，如图 5-9 所示。

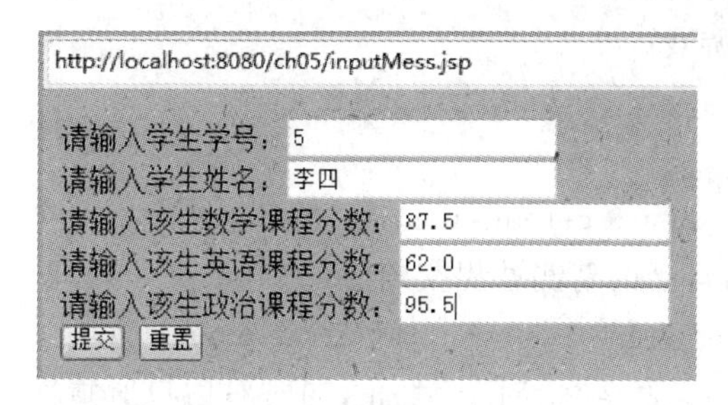

图 5-8 信息输入页面

图 5-9 信息显示页面

通过 HTTP 表单的参数的值设置 bean 的属性时，表单的参数的名字必须与 bean 属性的名字相同，服务器会根据名字自动匹配，类型自动转换。

由于客户可能通过表单提交汉字字符，可以采用 request.setCharacterEncoding("GBK")避免出现中文乱码。采用该方式避免中文乱码时，表单的提交方式一定是 post 的方式。

5.4.3 拓展训练

1. 在 JSP 中调用 JavaBean 时不会用到的标记是(　　)。

   A. <javabean>　　B. <jsp:useBean>

   C. <jsp:setProperty>　　D. <jsp:getProperty>

2. 在 J2EE 中，以下是有关 jsp:setProperty 和 jsp:getProperty 标记的描述，其中正确的是(　　)。

   A. <jsp:setProperty>和<jsp:getProperty>标记都必须在<jsp:useBean>的开始标记和结束标记之间

   B. 这两个标记的 name 属性的值必须和<jsp:useBean>标记的 id 属性的值相对应

C. 这两个标记的 name 属性的值可以和<jsp:userBean>标记的 id 属性的值不同

D. 以上均不对

3. 编写两个 JSP 页面：inputNumber. jsp 与 showResult. jsp。inputNumber. jsp 提供一个表单，用户可以通过表单输入两个数和四则运算符号提交给 showResult. jsp。用户提交表单后，JSP 页面将计算任务交给一个 bean 去完成，创建 bean 的源文件 Computer. java（在包 compute. add 中）。

5.5 JSP 与 JavaBean 结合实例

5.5.1 知识要点

通过前面的学习已经知道，在 JSP 页面中使用 JavaBean 可以将数据的处理代码从页面中分离出来，提高了代码的复用程度，方便了代码的维护。为了进一步掌握 JavaBean 的使用方法，本节给出了一个综合性的实例：猜数字游戏。

5.5.2 技能操作

编写两个 JSP 页面：getNumber. jsp 与 guess. jsp。getNumber. jsp 页面能够产生一个 1～100 之间的随机数，用户可以在文本框中给出自己的猜测并提交给 guess. jsp 页面处理；用户提交猜测后，guess. jsp 页面将验证猜测的任务交给一个 bean 即 GuessNumber. java 去完成，该 bean 能够提示用户"猜大了"或是"猜小了"直至"猜对了"为止，同时对用户所猜次数进行统计。

代码模板 GuessNumber. java 如下：

```
package china.dalian;
public class GuessNumber
{   int answer=0,                          //待猜测的整数
    guessNumber=0,                         //用户的猜测
    guessCount=0;                          //用户猜测的次数
    String result=null;
    boolean right=false;
    public void setAnswer(int n)
    {   answer=n;
        guessCount=0;
    }
   public int getAnswer()
    {   return answer;
    }
   public void setGuessNumber(int n)
    { guessNumber=n;
        guessCount++;
```

```
        if(guessNumber==answer)
        {   result="恭喜,猜对了";
            right=true;
        }
        else if(guessNumber>answer)
        {   result="猜大了";
              right=false;
        }
        else if(guessNumber<answer)
        {   result="猜小了";
              right=false;
        }
       }
    public int getGuessNumber()
     {   return guessNumber;
     }
    public int getGuessCount()
     {   return guessCount;
     }
    public String getResult()
     {   return result;
     }
    public boolean isRight()
     {   return right;
     }
}
```

代码模板 getNumber.jsp 如下:

```
<%@ page contentType="text/html;charset=GB2312" %>
<%@ page import="china.dalian.GuessNumber" %>
<html><body>
<% int n=(int)(Math.random() * 100)+1;%>
<jsp:useBean id="guess" class="china.dalian.GuessNumber" scope="session" />
<jsp:setProperty name= "guess" property="answer" value="<%=n%>" />
<p>随机给你一个 1 到 100 之间的数,请猜测这个数是多少?
<% String str=response.encodeRedirectURL("guess.jsp");
%>
<form action="<%=str%>" method=post>
<br>输入你的猜测:<input type=text name="guessNumber">
<input type=submit value="提交">
</form></body>
</html>
```

代码模板 guess.jsp 如下:

```
<%@ page contentType="text/html;charset=GB2312" %>
```

```
<%@ page import="china.dalian.GuessNumber" %>
<% String strGuess=response.encodeRedirectURL("guess.jsp"),
    strGetNumber=response.encodeRedirectURL("getNumber.jsp");
%>
<html><body>
<jsp:useBean id="guess" class="china.dalian.GuessNumber" scope="session" />
<jsp:setProperty name= "guess" property="guessNumber" param="guessNumber" />
<br><jsp:getProperty name= "guess" property="result" />,这是第
    <jsp:getProperty name="guess" property="guessCount" />猜.
    你给出的数是<jsp:getProperty name= "guess" property="guessNumber" />
<% if(guess.isRight()==false)
   {
%>    <form action="<%=strGuess%>" method=post>
      再输入你的猜测:<input type=text name="guessNumber">
      <input type=submit value="提交">
      </form>
<% }
%>
<br>
 <a href="<%=strGetNumber%>">回到首页,重新开始玩猜数字游戏!
 </a>
</body>
</html>
```

首先运行 getNumber.jsp 页面,如图 5-10 所示;当用户给出猜测数字"50"并单击"提交"按钮后,出现如图 5-11 所示的页面效果;直至最后将数字猜对,会出现如图 5-12 所示的页面效果。

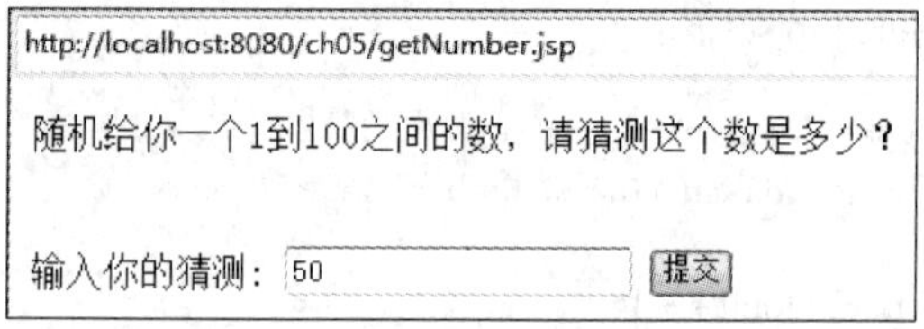

http://localhost:8080/ch05/getNumber.jsp

随机给你一个1到100之间的数,请猜测这个数是多少?

输入你的猜测: 50 提交

图 5-10　输入数字页面

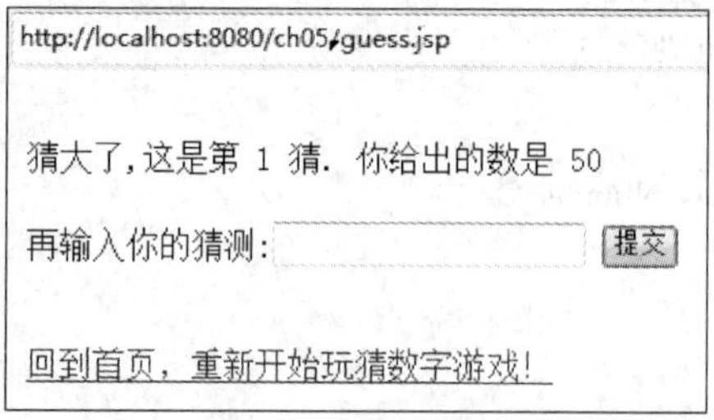

http://localhost:8080/ch05/guess.jsp

猜大了,这是第 1 猜. 你给出的数是 50

再输入你的猜测: 提交

回到首页,重新开始玩猜数字游戏!

图 5-11　猜测过程页面

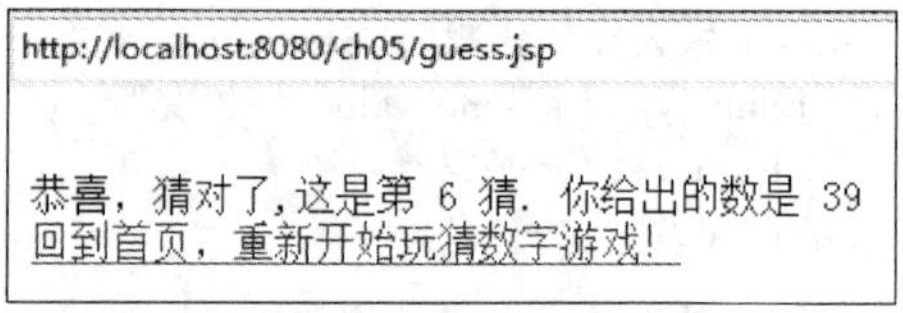

http://localhost:8080/ch05/guess.jsp

恭喜,猜对了,这是第 6 猜. 你给出的数是 39
回到首页,重新开始玩猜数字游戏!

图 5-12　猜对数字页面

5.5.3　拓展训练

运用本章所学的 JavaBean 相关内容，修改 4.2.3 节中第 7 题的简易计算器。具体要求如下：编写 JSP 页面 caculator.jsp，在该页面中用户可以输入两个运算数，可以选择加、减、乘、除运算符进行运算；页面将计算的任务提交给一个 bean 即 Calculator.java 去完成。页面运行效果如图 5-13 所示。

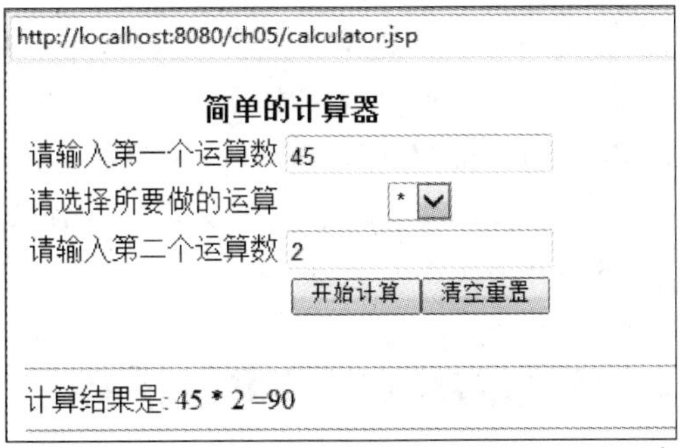

http://localhost:8080/ch05/calculator.jsp

简单的计算器

请输入第一个运算数 45

请选择所要做的运算 *

请输入第二个运算数 2

开始计算　清空重置

计算结果是: 45 * 2 =90

图 5-13　简易计算器

5.6　小　　结

- JavaBean 的实质是一种特殊的 Java 类，是一个可重复使用的软件组件，是遵循一定标准、用 Java 语言编写的一个类，该类的一个实例称为一个 JavaBean，简称 bean。
- JSP 和 JavaBean 技术的结合不仅可以实现数据的表示和处理分离，而且提高了代码的可重用、可维护性。
- JavaBean 的生命期限分为 page、request、session 和 application。

第 6 章

JSP 对数据库的访问

数据库在现在的 Web 应用中扮演着越来越重要的作用。如果没有数据库,很多重要的应用,像电子商务、搜索引擎等都不可能实现。本章主要介绍在 JSP 中如何访问关系数据库,如 Oracle、SQL Server、MySQL 和 Microsoft Access 等数据库。

本章将新建一个 Web 工程 ch06,本章例子中涉及的 Java 源文件保存在 ch06 的 src 中,涉及的 JSP 页面保存在 ch06 的 WebContent 中。

6.1 使用 JDBC-ODBC 桥接数据库

JSP 页面中访问数据库,首先要与数据库进行连接,通过连接向数据库发送指令,并获得返回的结果。JDBC 连接数据库通常有两种常用方式:建立 JDBC-ODBC 桥接器和加载纯 Java 驱动程序,本节主要介绍前一种,6.2 节将介绍第二种。

6.1.1 知识要点

JDBC(Java DataBase Connectivity)是用于运行 SQL 的解决方案,是 Java 运行平台核心类库中的一部分,它由一组标准接口与类组成。我们经常使用 JDBC 完成 3 件事:与指定的数据库建立连接;向已连接的数据库发送 SQL 命令;处理 SQL 命令返回的结果。

ODBC(Open DataBase Connectivity)是由 Microsoft 主导的数据库连接标准,提供了通用的数据库访问平台。但是,使用 ODBC 连接数据库的应用程序移植性较差,因为应用程序所在的计算机必须提供 ODBC。

使用 JDBC-ODBC 桥接器连接数据库的机制是:将连接数据库的相关信息提供给 JDBC-ODBC 驱动程序,然后转换成 JDBC 接口,供应用程序使用,而和数据库的连接是由 ODBC 完成。使用 JDBC-ODBC 桥接器连接数据库的示意图如图 6-1 所示。

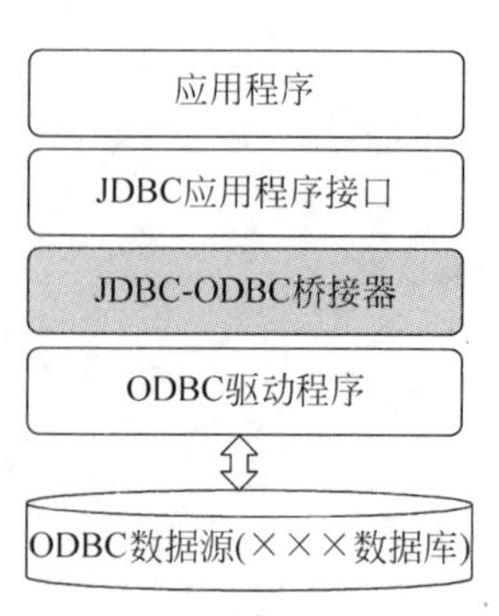

图 6-1 JDBC-ODBC 桥接器

使用 JDBC-ODBC 桥接器连接数据库有 3 个步骤:

(1) 建立 JDBC-ODBC 桥接器。

(2) 创建 ODBC 数据源。

(3) 和 ODBC 数据源指定的数据库建立连接。

6.1.2 技能操作

运用 JDBC-ODBC 桥接器连接数据库，具体任务如下：

- 创建待连接的 Microsoft Access 数据库。
- 建立 JDBC-ODBC 桥接器。
- 创建 ODBC 数据源。
- 和 ODBC 数据源指定的数据库建立连接。
- 在 JSP 页面中使用 JDBC-ODBC 桥接器连接数据库。

1. 创建待连接的 Microsoft Access 数据库

使用 Microsoft Access 2007 设计一个教师数据库 teacherDB 并保存到“D:\”盘根目录，该库中有一张教师信息表 teacherInfo，表的字段如图 6-2 所示。

teacherInfo

字段名称	数据类型	说明
teacherID	文本	教师编号
teacherName	文本	教师姓名
teacherLevel	文本	教师职称
teaherTime	文本	教师工作时间
teacherSalary	货币	教师工资

图 6-2　teacherInfo 表的说明

表中的数据如图 6-3 所示。

teacherInfo

teacherID	teacherName	teacherLevel	teacherTime	teacherSalary
1	赵晓晓	助教	2015/3/1	¥2,500.00
2	钱多多	讲师	2012/9/1	¥3,000.00
3	孙苗苗	副教授	2002/3/5	¥3,800.00
4	李大海	副教授	2000/3/1	¥4,000.00
5	周正	教授	1994/9/10	¥5,500.00

图 6-3　teacherInfo 表中的数据

2. 创建 ODBC 数据源

创建 ODBC 数据源时，必须保证计算机有 ODBC 系统，Windows 操作系统一般都带有 ODBC 系统。

1）打开 ODBC 数据源管理器

在 Windows 7 系统中，在“控制面板”中选择“系统和安全”→“管理工具”选项（有些 Windows XP 系统，需在“控制面板”中选择“性能和维护”→“管理工具”选项），找到“数据源(ODBC)”图标双击打开，出现如图 6-4 所示的界面。

2）为数据源选择驱动程序

在图 6-4 所示的界面选择“用户 DSN”选项卡，单击“添加”按钮，出现为新增的数据源选择驱动程序界面，如图 6-5 所示。因为我们要连接 Microsoft Access 2007 数据库，所以选择“Microsoft Access Driver(*.mdb, *.accdb)”，单击“完成”按钮。

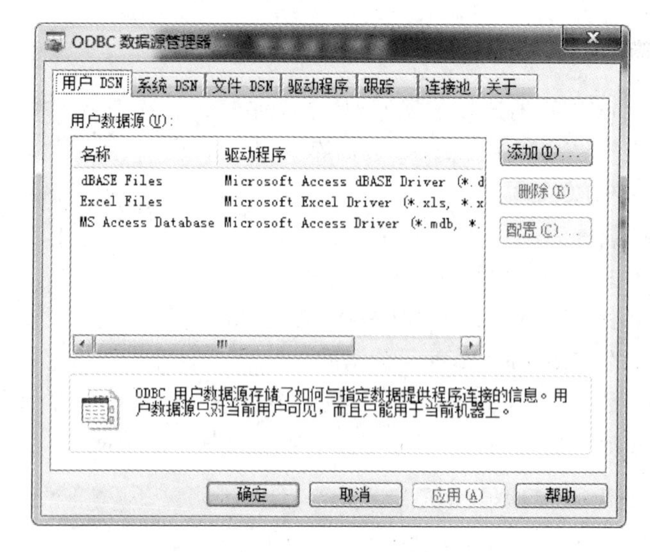

图 6-4　打开 ODBC 数据源管理器

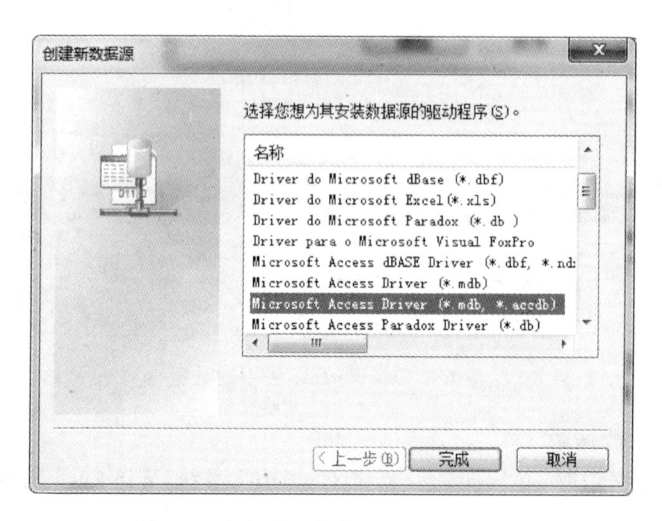

图 6-5　为新增的数据源选择驱动程序

3）为数据源起名并找到对应的数据库

在如图 6-5 所示的界面单击“完成”按钮出现设置数据源具体信息的对话框，如图 6-6 所示。在“数据源名”文本框中为数据源起个名字：teacher。在图 6-6 中单击“选择”按钮为 teacher 数据源选择数据库“D:\teacherDB.accdb”之后确定，如图 6-7 所示。

4）设置登录名和密码

回到如图 6-6 所示的界面，单击“高级”按钮，出现设置登录名与密码界面，如图 6-8 所示。这里的用户名是 tt，密码是 123。在如图 6-8 所示的界面单击“确定”按钮后，再单击如

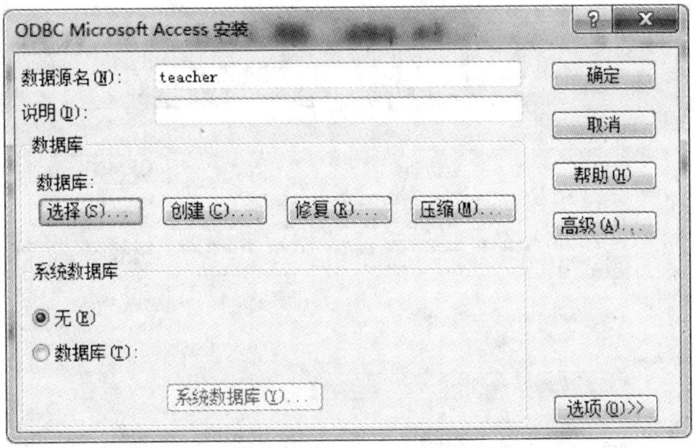

图 6-6　设置数据源的名字和选择数据库

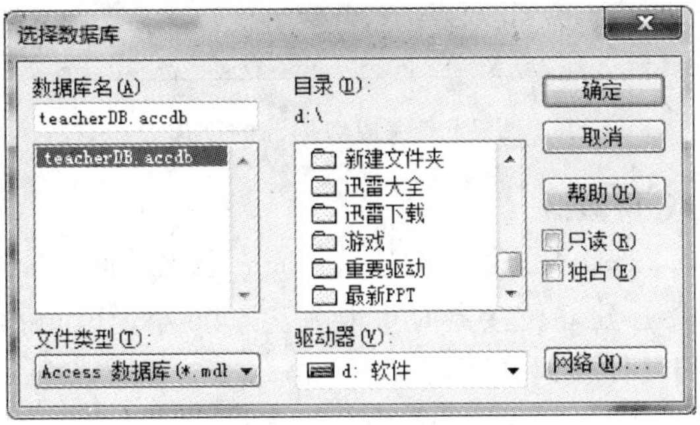

图 6-7　选择数据库

图 6-6 所示的界面中的"确定"按钮就创建了一个新的数据源：teacher，看到如图 6-9 所示的界面就表示数据已创建完成。

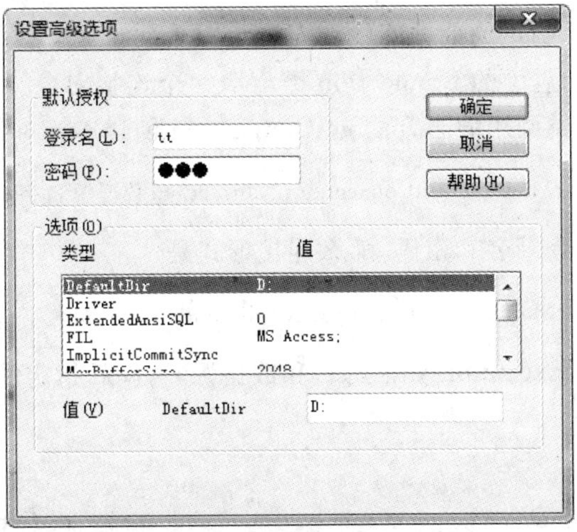

图 6-8　设置登录名和密码

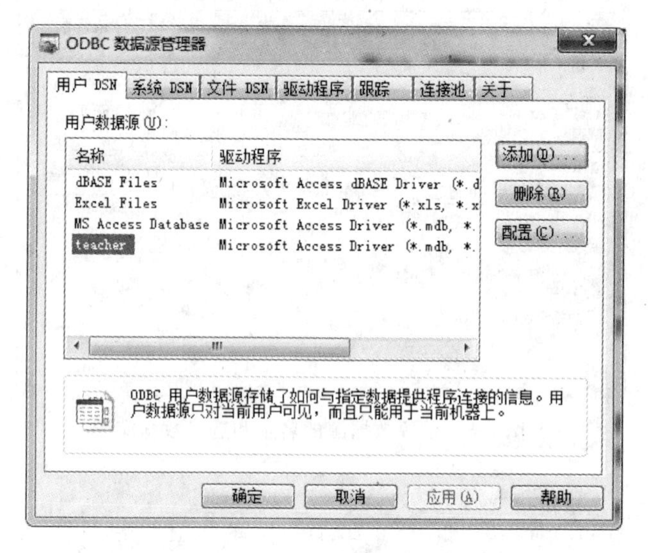

图 6-9 完成数据源的创建

3. 建立 JDBC-ODBC 桥接器

JDBC 通过 java.lang.Class 类的静态方法 forName 加载 sun.jdbc.odbc.JdbcOdbcDriver 类建立 JDBC-ODBC 桥接器。建立桥接器时可能发生 ClassNotFoundException 异常，必须捕获该异常，建立桥接器的具体代码如下：

```
try{
        Class.forName("sun.jdbc.odbc.JdbcOdbcDriver");
}catch(ClassNotFoundException e){
        e.printStackTrace();
}
```

4. 与 ODBC 数据源指定的数据库建立连接

首先，使用 java.sql 包中的 Connection 类声明一个连接对象 con，然后再使用 java.sql 包中的 DriverManager 类调用静态方法 getConnection 创建连接对象 con：

```
Connection con=DriverManager.getConnection("jdbc:odbc:数据源名字","登录名","密码");
```

如果没有给数据源设置登录名和密码，那么连接形式是：

```
Connection con=DriverManager.getConnection("jdbc:odbc:数据源名字","","");
```

建立连接时应捕获 SQLException 异常，如，和数据源 myGod 指定的数据库 goods.accdb 建立连接代码如下：

```
try{
        Connection con=DriverManager.getConnection("jdbc:odbc:myGod","firstDB","firstDB");
}catch(SQLException e){
```

```
        e.printStackTrace();
    }
```

5. 在 JSP 页面中使用 JDBC-ODBC 桥接器连接数据库

编写一个 JSP 页面 eg6_1.jsp，该页面中的 Java 程序片代码使用 JDBC-ODBC 桥接器连接到数据源 teacher，查询 teacherInfo 表中的全部记录。页面运行效果如图 6-10 所示。

http://localhost:8080/ch06/eg6_1.jsp

教师编号	教师姓名	教师职称	入职时间	教师工资
1	赵晓晓	助教	2015-03-01 00:00:00	2500.0000
2	钱多多	讲师	2012-09-01 00:00:00	3000.0000
3	孙苗苗	副教授	2002-03-05 00:00:00	3800.0000
4	李大海	副教授	2000-03-01 00:00:00	4000.0000
5	周正	教授	1994-09-10 00:00:00	5500.0000

图 6-10　使用 JDBC-ODBC 桥接器连接数据库

代码模板 eg6_1.jsp 如下：

```
<%@ page language="java" contentType="text/html; charset=GBK"
    pageEncoding="GBK"%>
<%@ page import="java.sql.*"%>
<html>
<head>
<title>eg6_1.jsp</title>
</head>
<body bgcolor="lightyellow">
    <%
        Connection con = null;
        Statement st = null;
        ResultSet rs = null;
        try {
            Class.forName("sun.jdbc.odbc.JdbcOdbcDriver");
        } catch (ClassNotFoundException e) {
            e.printStackTrace();
        }
        try {
            con = DriverManager.getConnection("jdbc:odbc:teacher","tt", "123");
            st=con.createStatement();
            rs=st.executeQuery("select * from teacherInfo");
            out.print("<table border=1>");
            out.print("<tr>");
                out.print("<th>教师编号</th>");
                out.print("<th>教师姓名</th>");
                out.print("<th>教师职称</th>");
```

```
                out.print("<th>入职时间</th>");
                out.print("<th>教师工资</th>");
            out.print("</tr>");
            while(rs.next()){
                out.print("<tr>");
                    out.print("<td>"+rs.getString(1)+"</td>");
                    out.print("<td>"+rs.getString(2)+"</td>");
                    out.print("<td>"+rs.getString(3)+"</td>");
                    out.print("<td>"+rs.getString(4)+"</td>");
                    out.print("<td>"+rs.getString(5)+"</td>");
                out.print("</tr>");
            }
            out.print("</table>");
        } catch (SQLException e) {
            e.printStackTrace();
        } finally{
            try{
                if(rs!=null){
                    rs.close();
                }
                if(st!=null){
                    st.close();
                }
                if(con!=null){
                    con.close();
                }
            }catch (SQLException e) {
                e.printStackTrace();
            }
        }
    %>
</body>
</html>
```

6.1.3 拓展训练

1. 当在 JSP 文件中要编写代码连接数据库时，应在 JSP 文件中加入以下哪个语句？(　　)

A. <jsp:include file="java.util.*"/>

B. <%@page import="java.sql.*" %>

C. <jsp:include page="java.lang.*"/>

D. <%@page import ="java.util.*"%>

2. Java 程序连接数据库常用的两种方式：建立 JDBC-ODBC 和加载纯（　　）驱动程序。

A. Oracle　　B. Java　　C. Java 数据库　　D. 以上都不对

3. JDBC 连接数据库常用的方式有哪些？

4. 参考 6.1.2 节中的主要内容，创建数据源 mySource，该数据源指定的数据库是 teacherDB.accdb。

编写一个 JSP 页面 practice6_1.jsp，该页面中的 Java 程序片代码使用 JDBC-ODBC 桥接器连接到数据源 mySource，查询 teacherInfo 表中工资高于 3000 元的教师信息（即 teacherSalary 字段值大于 3000 的全部记录）。页面运行效果如图 6-11 所示。

http://localhost:8080/ch06/practice6_1.jsp

教师编号	教师姓名	教师职称	入职时间	教师工资
3	孙苗苗	副教授	2002-03-05 00:00:00	3800.0000
4	李大海	副教授	2000-03-01 00:00:00	4000.0000
5	周正	教授	1994-09-10 00:00:00	5500.0000

图 6-11　practice6_1.jsp 页面运行效果

6.2　使用纯 Java 数据库驱动程序连接数据库

6.2.1　知识要点

应用程序除了使用 JDBC-ODBC 桥接器连接数据库外，还通常使用 JDBC API 调用本地的纯 Java 数据库驱动程序和相应的数据库建立连接，如图 6-12 所示。

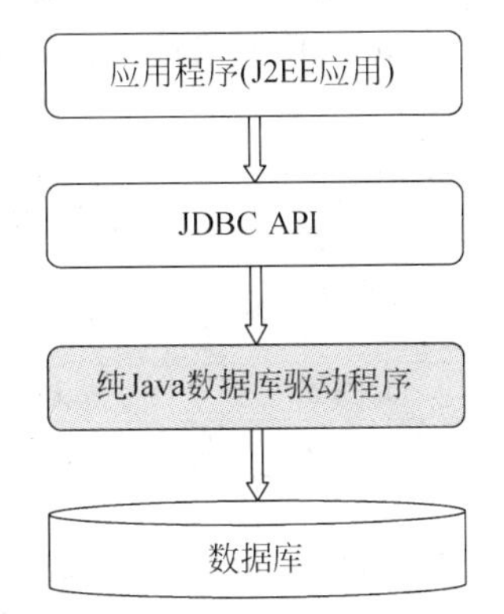

图 6-12　使用纯 Java 数据库驱动程序

使用纯 Java 数据库驱动程序连接数据库，需要经过两个步骤：

- 注册纯 Java 数据库驱动程序。
- 和指定的数据库建立连接。

下面以 Oracle10g 为例，讲解如何使用纯 Java 数据库驱动程序连接数据库的方法。

1. 注册纯 Java 数据库驱动程序

每种数据库都配有自己的纯 Java 数据库驱动程序。Oracle10g 的纯 Java 驱动程序一般位于数据库安装目录“\oracle\product\10.2.0\db_1\jdbc\lib”下，名为 classes12.jar。

为了连接 Oracle10g 数据库，可以将 classes12.jar 文件复制到 Web 应用程序的/WEB-

INF/lib 文件夹中。然后,通过 java. lang. Class 类的 forName(),动态注册 Oracle10g 的纯 Java 驱动程序,代码如下:

```
try {
        Class.forName("oracle.jdbc.driver.OracleDriver");
} catch (ClassNotFoundException e) {
        e.printStackTrace();
}
```

2. 和指定的数据库建立连接

和 Oracle 数据库建立连接的代码如下:

```
try {
    Connection con=DriverManager.getConnection("jdbc:oracle:thin:@主机:端口号:数据库名",
            "用户名","密码");
} catch (SQLException e) {
    e.printStackTrace();
}
```

其中,主机是安装 Oracle 服务器的 IP 地址,如果是本机则为 localhost; Oracle 默认端口号为 1521; Oracle 默认数据库名为 orcl; 用户名和密码是访问 Oracle 服务器的用户权限。

本章中后续的例子(除 6.5 节与 6.7 节)均采用纯 Java 数据库驱动程序连接 Oracle10g。

6.2.2 技能操作

使用纯 Java 数据库驱动程序连接数据库,具体任务如下:

- 利用 Oracle 建立商品数据库 goods,在该数据库中建立商品信息表 goosInfo, goosInfo 表的字段如表 6-1 所示。

表 6-1 商品信息表

字段名	类型	长度	限制	备注
goodsId	number	(4)	非空,主键	商品编号
goodsName	varchar	(50)	非空	商品名称
goodsPrice	number	(7,2)	非空	商品价格
goodsType	varchar	(10)	非空	商品类型

- 编写一个 JSP 页面 eg6_2.jsp,该页面中的 Java 程序片代码使用纯 Java 驱动程序连接 Oracle 数据库,查询 goodsInfo 表中的全部记录。

创建 goodsInfo 表的 SQL 语句如下:

```
create table goodsinfo (
    goodsId number(4) not null,
```

```
    goodsName varchar(50) not null,
    goodsPrice number(7,2) not null,
    goodsType varchar(10) not null,
    constraint pk_goodsinfo primary key (goodsId)
);
insert into goodsinfo values(1,'牙膏',12,'日用品');
insert into goodsinfo values(2,'冰箱',2500,'电器');
insert into goodsinfo values(3,'蛋糕',28,'食品');
insert into goodsinfo values(4,'苹果',48,'水果');
insert into goodsinfo values(5,'上衣',1800,'服装');
insert into goodsinfo values(6,'书包',78,'文具');
commit;
```

页面运行效果如图 6-13 所示。

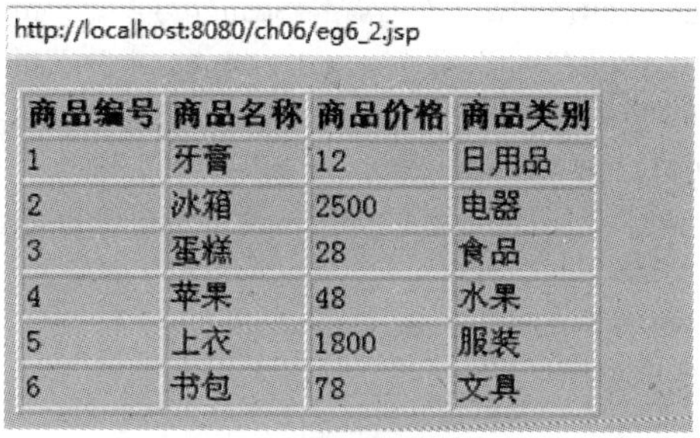

http://localhost:8080/ch06/eg6_2.jsp

商品编号	商品名称	商品价格	商品类别
1	牙膏	12	日用品
2	冰箱	2500	电器
3	蛋糕	28	食品
4	苹果	48	水果
5	上衣	1800	服装
6	书包	78	文具

图 6-13　使用纯 Java 驱动程序连接 Oracle 数据库

代码模板 eg6_2.jsp 如下：

```
<%@ page language="java" contentType="text/html; charset=GBK"
    pageEncoding="GBK"%>
<%@ page import="java.sql.*"%>
<html>
<head>
<title>eg6_2.jsp</title>
</head>
<body bgcolor="LightPink">
    <%
        Connection con = null;
        Statement st = null;
        ResultSet rs = null;
        try {
          Class.forName("oracle.jdbc.driver.OracleDriver");//注册 Oracle 的纯 Java 驱动程序
        } catch (ClassNotFoundException e) {
            e.printStackTrace();
        }
        try {
```

```
            con=DriverManager.getConnection("jdbc:oracle:thin:@localhost:1521:orcl",
                    "system","system");
            st=con.createStatement();
            rs=st.executeQuery("select * from goodsInfo");
            out.print("<table border=1>");
            out.print("<tr>");
                out.print("<th>商品编号</th>");
                out.print("<th>商品名称</th>");
                out.print("<th>商品价格</th>");
                out.print("<th>商品类别</th>");
            out.print("</tr>");
            while(rs.next()){
                out.print("<tr>");
                    out.print("<td>"+rs.getString(1)+"</td>");
                    out.print("<td>"+rs.getString(2)+"</td>");
                    out.print("<td>"+rs.getString(3)+"</td>");
                    out.print("<td>"+rs.getString(4)+"</td>");
                out.print("</tr>");
            }
            out.print("</table>");
        } catch (SQLException e) {
            e.printStackTrace();
        }finally{
            try{
                if(rs!=null){
                    rs.close();
                }
                if(st!=null){
                    st.close();
                }
                if(con!=null){
                    con.close();
                }
            }catch (SQLException e) {
                e.printStackTrace();
            }
        }
    %>
</body>
</html>
```

应用程序连接 Oracle 数据库时，必须事先启动 Oracle 服务器的 OracleServiceORCL 和 OracleOraDb10g_home1TNSListener 两个服务，否则会抛出连接异常。

从任务中可以看出编写程序访问数据库需要有以下几个步骤：

(1) 导入java.sql包。

所有与数据库有关的对象和方法都在java.sql包中,包java.sql包含了用Java操作关系数据库的类和接口。因此在使用JDBC的程序中必须要加入import java.sql.*。

(2) 加载驱动程序。

在该任务中使用了Class类(java.lang包)中的方法forName,来装入该驱动程序的类定义oracle.jdbc.driver.OracleDriver,从而创建了该驱动程序的一个实例。

(3) 连接数据库。

完成上述操作后,就可以连接一个特定的数据库了。这需要创建Connection类的一个实例,并使用DriverManager的方法getConnection来尝试建立用url指定的数据库的连接。代码如下:

```
con=DriverManager.getConnection("jdbc:oracle:thin:@localhost:1521:orcl","system","system");
```

(4) 访问数据库。

访问数据库时,需要先用Connection类的createStatement方法从指定的数据库连接得到一个Statement的实例,然后用这个实例的executeQuery方法来执行一条SQL语句。代码如下:

```
st=con.createStatement();
rs=st.executeQuery("select * from goodsInfo");
```

(5) 处理返回的结果集。

ResultSet对象是JDBC中比较重要的一个对象,几乎所有的查询操作都将数据作为ResultSet对象返回。处理结果集ResultSet对象的代码如下:

```
while(rs.next()){
    out.print("<tr>");
    out.print("<td>"+rs.getString(1)+"</td>");
    out.print("<td>"+rs.getString(2)+"</td>");
    out.print("<td>"+rs.getString(3)+"</td>");
    out.print("<td>"+rs.getString(4)+"</td>");
    out.print("</tr>");
}
```

(6) 关闭数据库连接,释放资源。

对数据库的操作完成之后,要及时关闭ResultSet对象、Statement对象和数据库连接对象Connection,从而释放占用的资源,这就要用到close方法。代码如下:

```
rs.close();
st.close();
con.close();
```

关闭的顺序从先到后依次为ResultSet对象、Statement对象和Connection对象。

6.2.3 拓展训练

使用纯 Java 数据库驱动程序访问数据库时，都有哪些步骤？

编写一个 JSP 页面 practice6_2.jsp，该页面中的 Java 程序片代码使用纯 Java 驱动程序连接 Oracle 数据库，查询 goodsInfo 表中 goodsPrice 字段值大于 10 并小于 50 的全部记录。页面运行效果如图 6-14 所示。

http://localhost:8080/ch06/practice6_2.jsp

商品编号	商品名称	商品价格	商品类别
1	牙膏	12	日用品
3	蛋糕	28	食品
4	苹果	48	水果

图 6-14　practice6_2.jsp 页面运行效果

6.3 使用 Statement 和 ResultSet 操作数据

6.3.1 知识要点

与数据库建立连接之后，若接下来要执行 SQL 语句，需要进行以下几个步骤。

1. 创建 Statement 对象

Statement 对象代表一条发送到数据库执行的 SQL 语句。由已创建的 Connection 对象 con 调用 createStatement()方法来创建 Statement 对象，代码如下：

```
Statement smt=con.createStatement();
```

2. 执行 SQL 语句

创建 Statement 对象之后，可以使用 Statement 对象调用 executeUpdate(String sql)、executeQuery(String sql)等方法来执行 SQL 语句。

executeUpdate(String sql)方法主要用于执行 INSERT、UPDATE 或 DELETE 语句以及 SQL DDL 语句，例如 CREATE TABLE 和 DROP TABLE。该方法返回一个整数(代表被更新的行数)，对于 CREATE TABLE 和 DROP TABLE 等不操作行的指令，返回零。

executeQuery(String sql)方法则是用于执行 SELECT 等查询数据库的 SQL 语句，该方法返回 ResultSet 对象，代表查询的结果。

3. 处理返回的 ResultSet 对象

ResultSet 对象是 executeQuery(String sql)方法的返回值，被称为结果集，它代表符合 SQL 语句条件的所有行。ResultSet 对象调用 next()方法移动到下一个数据行(顺序查询)，当数据行存在时，next()方法返回 true，否则返回 false。获得一行数据后，ResultSet 对象可以使用 getXxx 方法获得字段值，getXxx 方法都提供依字段名称取得数据，或是依字段顺序取得数据的方法。

6.3.2　技能操作

使用 Statement 与 ResultSet 对象对数据库进行增删改查，具体任务如下：编写两个 JSP 页面——addGoods.jsp 和 showAllGoods.jsp。用户可以在 addGoods.jsp 页面中输入信息后，单击"添加"按钮把信息添加到 goodsInfo 表中。然后，在 showAllGoods.jsp 页面中显示所有商品信息。在该任务中需要编写一个 bean(GoodsBean.java)用来实现添加和查询记录。页面运行效果如图 6-15 和图 6-16 所示。

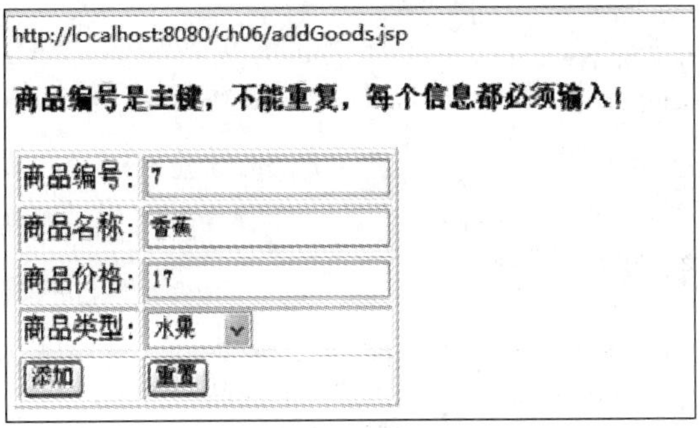

http://localhost:8080/ch06/addGoods.jsp

商品编号是主键，不能重复，每个信息都必须输入！

商品编号：	7
商品名称：	香蕉
商品价格：	17
商品类型：	水果
添加	重置

图 6-15　添加记录

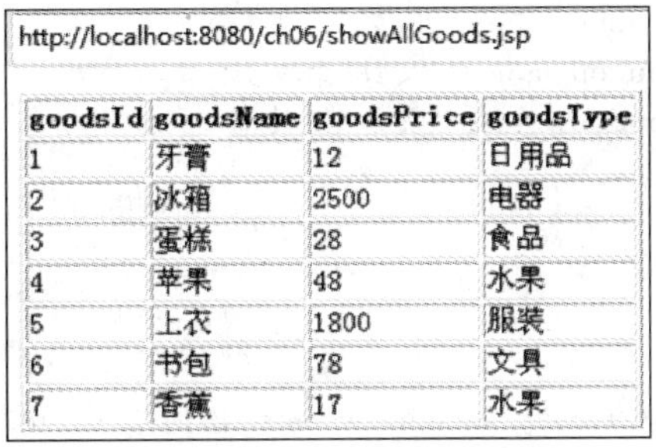

http://localhost:8080/ch06/showAllGoods.jsp

goodsId	goodsName	goodsPrice	goodsType
1	牙膏	12	日用品
2	冰箱	2500	电器
3	蛋糕	28	食品
4	苹果	48	水果
5	上衣	1800	服装
6	书包	78	文具
7	香蕉	17	水果

图 6-16　查询记录

代码模板 addGoods.jsp 如下：

```
<%@ page language="java" contentType="text/html; charset=GBK"pageEncoding="GBK"%>
<html>
<head>
<title>addGoods.jsp</title>
</head>
<body>
    <h4>商品编号是主键,不能重复,每个信息都必须输入!</h4>
```

```
        <form action="showAllGoods.jsp" method="post">
        <table border="1">
            <tr>
                <td>商品编号:</td>
                <td><input type="text" name="goodsId"/></td>
            </tr>

            <tr>
                <td>商品名称:</td>
                <td><input type="text" name="goodsName"/></td>
            </tr>

            <tr>
                <td>商品价格:</td>
                <td><input type="text" name="goodsPrice"/></td>
            </tr>

            <tr>
                <td>商品类型:</td>
                <td>
                    <select name="goodsType">
                        <option value="日用品">日用品
                        <option value="电器">电器
                        <option value="食品">食品
                        <option value="水果">水果
                        <option value="服装">服装
                        <option value="文具">文具
                        <option value="其他">其他
                    </select>
                </td>
            </tr>

            <tr>
                <td><input type="submit" value="添加"></td>
                <td><input type="reset" value="重置"></td>
            </tr>
        </table>
        </form>
    </body>
    </html>
```

代码模板 showAllGoods.jsp 如下：

```
<%@ page language="java" contentType="text/html; charset=GBK"pageEncoding="GBK"%>
<%@ page import="bean.GoodsBean" %>
<html>
```

```
<head>
<title>showAllGoods.jsp</title>
</head>
<body>
    <%
        request.setCharacterEncoding("GBK");
    %>
    <jsp:useBean id="goods" class="bean.GoodsBean" scope="page"></jsp:useBean>
    <jsp:setProperty property=" * " name="goods"/>
    <%
        goods.addGoods();//添加商品
    %>
    <jsp:getProperty property="queryResult" name="goods"/><!-- 获得查询结果 -->
</body>
</html>
```

代码模板 GoodsBean.java 如下：

```
package bean;
import java.sql. * ;
public class GoodsBean {
    int goodsId;
    String goodsName;
    double goodsPrice;
    String goodsType;
    StringBuffer queryResult;                        //查询结果
    public GoodsBean(){

    }
    public int getGoodsId() {
        return goodsId;
    }
    public void setGoodsId(int goodsId) {
        this.goodsId = goodsId;
    }
    public String getGoodsName() {
        return goodsName;
    }
    public void setGoodsName(String goodsName) {
        this.goodsName = goodsName;
    }
    public double getGoodsPrice() {
        return goodsPrice;
    }
    public void setGoodsPrice(double goodsPrice) {
        this.goodsPrice = goodsPrice;
```

```
    }
    public String getGoodsType() {
        return goodsType;
    }
    public void setGoodsType(String goodsType) {
        this.goodsType = goodsType;
    }
    //添加记录
    public void addGoods(){
        Connection con = null;
        Statement st = null;
        try {
            Class.forName("oracle.jdbc.driver.OracleDriver");
        } catch (ClassNotFoundException e) {
            e.printStackTrace();
        }
        try {
            con=DriverManager.getConnection("jdbc:oracle:thin:@localhost:1521:orcl",
                    "system","system");
            st=con.createStatement();                //创建 Statement 对象
            String addSql="insert into goodsInfo
                values("+goodsId+",'"+goodsName+"',"+goodsPrice+",'"+goodsType+"')";
            st.executeUpdate(addSql);                //执行 insert 语句
        }catch (SQLException e) {
            e.printStackTrace();
        }finally{
            try{
                if(st!=null){
                    st.close();
                }
                if(con!=null){
                    con.close();
                }
            }catch (SQLException e) {
                e.printStackTrace();
            }
        }
    }
    //查询记录
    public StringBuffer getQueryResult(){
        queryResult=new StringBuffer();
        Connection con = null;
        Statement st = null;
        ResultSet rs=null;
        try {
```

```
            Class.forName("oracle.jdbc.driver.OracleDriver");
        } catch (ClassNotFoundException e) {
            e.printStackTrace();
        }
        try {
            con=DriverManager.getConnection("jdbc:oracle:thin:@localhost:1521:orcl",
                    "system","system");
            st=con.createStatement();              //创建 Statement 对象
            String selectSql="select * from goodsInfo";
            rs=st.executeQuery(selectSql);         //执行 select 语句
            queryResult.append("<table border=1>");
                queryResult.append("<tr>");
                    queryResult.append("<th>goodsId</th>");
                    queryResult.append("<th>goodsName</th>");
                    queryResult.append("<th>goodsPrice</th>");
                    queryResult.append("<th>goodsType</th>");
                queryResult.append("</tr>");
            while(rs.next()){
                queryResult.append("<tr>");
                    queryResult.append("<td>"+rs.getString(1)+"</td>");
                    queryResult.append("<td>"+rs.getString(2)+"</td>");
                    queryResult.append("<td>"+rs.getString(3)+"</td>");
                    queryResult.append("<td>"+rs.getString(4)+"</td>");
                queryResult.append("</tr>");
            }
            queryResult.append("</table>");
        }catch (SQLException e) {
            e.printStackTrace();
        }finally{
            try{
                if(rs!=null){
                    rs.close();
                }
                if(st!=null){
                    st.close();
                }
                if(con!=null){
                    con.close();
                }
            }catch (SQLException e) {
                e.printStackTrace();
            }
        }
        return queryResult;
    }
}
```

ResultSet 对象自动维护指向其当前数据行的游标。每调用一次 next()方法，游标向下移动一行。最初它位于结果集的第一行之前，因此第一次调用 next()，将把游标置于第一行上，使它成为当前行。随着每次调用 next()，导致游标向下移动一行，按照从上至下的次序获取 ResultSet 行，实现顺序查询。

ResultSet 对象包含 SQL 语句的执行结果。它通过一套 get 方法对这些行中数据的访问，即使用 getXxx 方法获得数据。get 方法很多如表 6-2 所示，究竟用哪一个 getXxx()方法，由列的数据类型来决定。

通过表 6-2 列举的方法不难看出，在前面的例子中，也可以将 rs. getString(1)替换成 rs. getString("goodsId")，将 rs. getString(2)替换成 rs. getString("goodsName")，将 rs . getString(3)替换成 rs. getString("goodsPrice")，将 rs. getString(4)替换成 rs. getString ("goodsType")。

表 6-2　Result 对象的 getXxx()方法

返回类型	方法名称	返回类型	方法名称
byte	getByte(int columnIndex)	String	getByte(String columName)
date	getDate(int columnIndex)	date	getDate(String columName)
double	getDouble(int columnIndex)	double	getDouble(String columName)
float	getFloat(int columnIndex)	float	getFloat(String columName)
int	getInt(int columnIndex)	int	getInt(String columName)
String	getString(int columnIndex)	String	getString(String columName)

使用 getXxx 方法时，还需要注意以下两点：

(1) 无论列是何种数据类型，总可以使用 getString(int columnIndex)或 getString (String columnName)方法获得列值的字符串表示。

(2) 如果使用 getString(int columnIndex)方法查看一行记录时，不允许颠倒顺序，例如，不允许：

```
rs.getString(2);
rs.getString(1);
```

6.3.3　拓展训练

1. 从“员工”表的“姓名”字段找出名字包含“玛丽”的人，下面的 select 语句中正确的是(　　)。

A. select * from 员工 where 姓名 like '%玛丽%'

B. select * from 员工 where 姓名＝ '%玛丽_'

C. select * from 员工 where 姓名 like '_玛丽%'

D. select * from 员工 where 姓名＝ '_玛丽_'

2. 下面(　　)不是 ResultSet 接口的方法。

A. next()　　B. getString()　　C. back()　　D. getInt()

3. 编写两个 JSP 页面：inputQuery.jsp 和 showGoods.jsp。用户可以在 inputQuery.jsp 页面输入查询条件后，单击"查询"按钮。然后，在 showGoods.jsp 页面中显示符合查询条件的商品信息。在本节任务的 bean(GoodsBean.java)中添加一个方法 getQueryResultBy()实现该题的条件查询功能。页面运行效果如图 6-17 和图 6-18 所示。

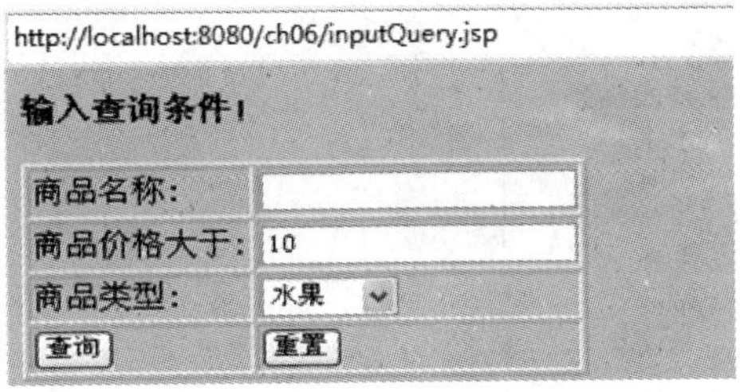

http://localhost:8080/ch06/inputQuery.jsp

输入查询条件！

商品名称：

商品价格大于：10

商品类型：水果

查询　重置

图 6-17　输入条件

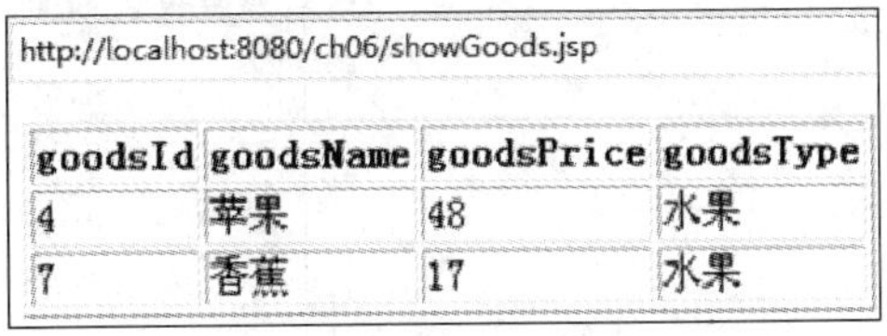

http://localhost:8080/ch06/showGoods.jsp

goodsId	goodsName	goodsPrice	goodsType
4	苹果	48	水果
7	香蕉	17	水果

图 6-18　符合查询条件的记录

6.4　游动查询

6.4.1　知识要点

有时候需要结果集的游标前后移动，这时可使用滚动结果集。为了获得滚动结果集，必须先用下面方法得到一个 Statement 对象：

```
Statement st=con.createStatement(int type, int concurrency);
```

根据 type 和 concurrency 的取值，当执行 ResultSet rs=st.executeQuery(String sql)时，会返回不同类型的结果集。

type 的取值决定滚动方式，它可以取如下值：

- ResultSet.TYPE_FORWORD_ONLY——表示结果集只能向下滚动。
- ResultSet.TYPE_SCROLL_INSENSITIVE——表示结果集可以上下滚动，当数据库变化时，结果集不变。
- ResultSet.TYPE_SCROLL_SENSITIVE——表示结果集可以上下滚动，当数据库变化时，结果集同步改变。

concurrency 取值表示是否可以用结果集更新数据库，它的取值是：

- ResultSet.CONCUR_READ_ONLY——表示不能用结果集更新数据库的表。
- ResultSet.CONCUR_UPDATETABLE——表示能用结果集更新数据库的表。

6.4.2 技能操作

使用滚动结果集进行游动查询,具体任务如下:编写一个 JSP 页面 randomQuery.jsp,查询 goodsInfo 表中的全部记录,并将结果逆序输出,最后单独输出第 4 条记录。运行结果如图 6-19 所示。

http://localhost:8080/ch06/randomQuery.jsp

该表共有6条记录
现在逆序输出记录:

商品编号	商品名称	商品价格	商品类别
7	香蕉	17	水果
6	书包	78	文具
5	上衣	1800	服装
4	苹果	48	水果
2	冰箱	2500	电器
1	牙膏	12	日用品

单独输出第4条记录
5 上衣 1800 服装

图 6-19 随机查询记录

代码模板 randomQuery.jsp 如下:

```
<%@ page language="java" contentType="text/html; charset=GBK"pageEncoding="GBK"%>
<%@ page import="java.sql.*"%>
<html>
<head>
<title>randomQuery.jsp</title>
</head>
<body>
    <%
        Connection con = null;
        Statement st = null;
        ResultSet rs = null;
        try {
            Class.forName("oracle.jdbc.driver.OracleDriver");
        } catch (ClassNotFoundException e) {
            e.printStackTrace();
        }
        try {
            con = DriverManager.getConnection("jdbc:oracle:thin:@localhost:1521:orcl",
"system","system");
            //创建 st 对象,该对象获得的结果集和数据库同步变化,但不能用结果集更新表
            st = con.createStatement(ResultSet.TYPE_SCROLL_SENSITIVE,
                    ResultSet.CONCUR_READ_ONLY);   //返回可滚动的结果集
```

```
        rs = st.executeQuery("SELECT * FROM goodsInfo");
        //将游标移动到最后一行
        rs.last();
        //获取最后一行的行号
        int lownumber = rs.getRow();
        out.print("该表共有" + lownumber + "条记录");
        out.print("<BR>现在逆序输出记录: ");
        out.print("<Table Border=1>");
        out.print("<TR>");
        out.print("<TH>商品编号</TH>");
        out.print("<TH>商品名称</TH>");
        out.print("<TH>商品价格</TH>");
        out.print("<TH>商品类别</TH>");
        out.print("</TR>");
        //为了逆序输出记录,需将游标移动到最后一行之后
        rs.afterLast();
        while (rs.previous()) {
            out.print("<TR>");
            out.print("<TD >" + rs.getString(1) + "</TD>");
            out.print("<TD >" + rs.getString(2) + "</TD>");
            out.print("<TD >" + rs.getString(3) + "</TD>");
            out.print("<TD >" + rs.getString(4) + "</TD>");
            out.print("</TR>");
        }
        out.print("</Table>");
        out.print("<BR>单独输出第 4 条记录<BR>");
        rs.absolute(4);
        out.print(rs.getString(1) + " ");
        out.print(rs.getString(2) + " ");
        out.print(rs.getString(3) + " ");
        out.print(rs.getString(4));
    } catch (SQLException e) {
        e.printStackTrace();
    } finally {
        try {
            if (rs != null) {
                rs.close();
            }
            if (st != null) {
                st.close();
            }
            if (con != null) {
                con.close();
```

```
                }
            } catch (SQLException e) {
                e.printStackTrace();
            }
        }
    %>
</body>
</html>
```

游动查询经常用到 ResultSet 类的下述方法：

- public boolean absolute(int row)——将游标移到参数 row 指定的行。如果 row 取负值，就是倒数的行，如－1 表示最后一行。当移到最后一行之后或第一行之前时，该方法返回 false。
- public void afterLast()——将游标移到结果集的最后一行之后。
- public void beforeFirst()——将游标移到结果集的第一行之前。
- public void first()——将游标移到结果集的第一行。
- public int getRow()——得到当前游标所指定的行号，如果没有行，则返回 0。
- public boolean isAfterLast()——判断游标是不是在结果集的最后一行之后。
- public boolean isBeforeFirst()——判断游标是不是在结果集的第一行之前。
- public void last()——将游标移到结果集的最后一行。
- public boolean previous()——将游标向上移动(和 next 方法相反)，当移动结果集的第一行之前时返回 false。

6.4.3 拓展训练

1. 下述选项中不属于 JDBC 基本功能的是(　　)。

   A. 与数据库建立连接　　B. 提交 SQL 语句

   C. 处理查询结果　　D. 数据库维护管理

2. 编写一个 JSP 页面 practice6_4.jsp，查询 goodsInfo 表中的记录，并逆序输出偶数行的记录。运行结果如图 6-20 所示。

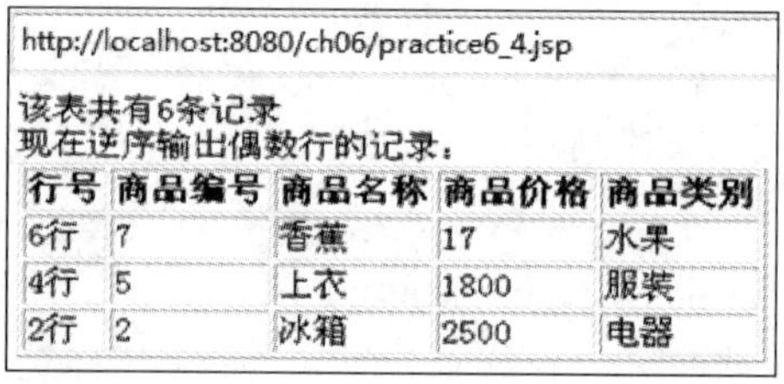

http://localhost:8080/ch06/practice6_4.jsp

该表共有6条记录
现在逆序输出偶数行的记录：

行号	商品编号	商品名称	商品价格	商品类别
6行	7	香蕉	17	水果
4行	5	上衣	1800	服装
2行	2	冰箱	2500	电器

图 6-20　逆序输出偶数行的记录

6.5 使用 PreparedStatement 操作数据

与 Statement 语句一样，PreparedStatement 也同样可以完成向数据库发送 SQL 语句，获取数据库操作结果的功能。但是 Statement 对象在每次执行 SQL 语句时都将该语句传送给数据库，然后数据库解释器负责将 SQL 语句转换成内部命令，并执行该命令，完成相应的数据库操作。基于这种机制，每次向数据库发送一条 SQL 语句时，都要先转化成内部命令，如果不断地执行程序，就会加重解释器的负担，影响执行的速度。

而 PreparedStatement 对象将 SQL 语句传给数据库进行预编译，以后需要执行同一条语句时就不需要再重新编译，直接执行就可以了，这样就大大提高了数据库的执行速度。

6.5.1 知识要点

可以使用 Connection 的对象 con 调用 prepareStatement(String sql)方法对参数 sql 指定的 SQL 语句进行预先编译，生成数据库的底层命令，并将该命令封装在 PreparedStatement 对象中。对于 SQL 语句中会变动的部分，可以使用通配符"?"代替。例如：

```
PreparedStatement ps=con.prepareStatement("insert into goodsInfo values(?,?,?,?)");
```

然后使用对应的 setXxx(int parameterIndex, xxx value)方法指定"?"代表的值，其中参数 parameterIndex 用来表示 SQL 语句中从左到右的第 parameterIndex 个通配符号，value 代表该通配符所代表的具体值。例如：

```
ps.setInt(1,9);
ps.setString(2, "手机");
ps.setDouble(3, 1900.8);
ps.setString(4, "通信");
```

若要让 SQL 语句执行生效，需使用 PreparedStatement 的对象 ps 调用 executeUpdate()方法。如果是查询的话，ps 就调用 executeQuery()方法，并返回 ResultSet 对象。

6.5.2 技能操作

使用 PreparedStatement 预处理语句对象操作数据库中的表。具体任务如下：

编写两个 JSP 页面——inputPrepareGoods.jsp 和 showPrepareGoods.jsp。用户可以在 inputPrepareGoods.jsp 页面中输入信息后，单击"添加"按钮把信息添加到 goodsInfo 表中。然后，在 showPrepareGoods.jsp 页面中显示所有商品信息。在该任务中需要编写一个 bean(UsePrepare.java)，bean 中使用预处理语句向 goodsInfo 表中添加记录。页面运行效果如图 6-21 和图 6-22 所示。

http://localhost:8080/ch06/inputPrepareGoods.jsp

商品编号是主键，不能重复，每个信息都必须输入!

商品编号:	10
商品名称:	豆浆机
商品价格:	299.99
商品类型:	电器
添加	重置

图 6-21　使用预处理添加商品

http://localhost:8080/ch06/showPrepareGoods.jsp

goodsId	goodsName	goodsPrice	goodsType
1	牙膏	12	日用品
2	冰箱	2500	电器
4	苹果	48	水果
5	上衣	1800	服装
6	书包	78	文具
7	香蕉	17	水果
10	豆浆机	299.99	电器

图 6-22　使用预处理查询商品

代码模板 inputPrepareGoods.jsp 如下：

```
<%@ page language="java" contentType="text/html; charset=GBK" pageEncoding=
"GBK"%>
<html>
<head>
<title>使用预处理语句</title>
</head>
<body bgcolor="LightYellow">
    <h4>商品编号是主键,不能重复,每个信息都必须输入!</h4>
    <form action="showPrepareGoods.jsp" method="post">
    <table border="1">
        <tr>
            <td>商品编号:</td>
            <td><input type="text" name="goodsId"/></td>
        </tr>

        <tr>
            <td>商品名称:</td>
            <td><input type="text" name="goodsName"/></td>
        </tr>

        <tr>
            <td>商品价格:</td>
            <td><input type="text" name="goodsPrice"/></td>
        </tr>

        <tr>
            <td>商品类型:</td>
            <td>
                <select name="goodsType">
                    <option value="日用品">日用品
                    <option value="电器">电器
                    <option value="食品">食品
```

```
                <option value="水果">水果
                <option value="服装">服装
                <option value="文具">文具
                <option value="其他">其他
            </select>
        </td>
    </tr>

    <tr>
        <td><input type="submit" value="添加"></td>
        <td><input type="reset" value="重置"></td>
    </tr>
  </table>
  </form>
</body>
</html>
```

代码模板 showPrepareGoods.jsp 如下：

```
<%@ page language="java" contentType="text/html; charset=GBK"pageEncoding="GBK"%>
<%@ page import="bean.UsePrepare" %>
<html>
<head>
<title>使用预处理语句</title>
</head>
<body>
    <%
        request.setCharacterEncoding("GBK");
    %>
    <jsp:useBean id="prepareGoods" class="bean.UsePrepare" scope="page"></jsp:
useBean>
    <jsp:setProperty property="*" name="prepareGoods"/>
    <%
        prepareGoods.addGoods();                //添加商品
    %>
    <jsp:getProperty property="queryResult" name="prepareGoods"/><!-- 获得查询结果 -->
</body>
</html>
```

代码模板 UsePrepare.java 如下：

```
package bean;
import java.sql.*;
public class UsePrepare {
    int goodsId;
    String goodsName;
```

```
double goodsPrice;
String goodsType;
StringBuffer queryResult;                           //查询所有商品结果
public UsePrepare(){

}
public int getGoodsId() {
    return goodsId;
}
public void setGoodsId(int goodsId) {
    this.goodsId = goodsId;
}
public String getGoodsName() {
    return goodsName;
}
public void setGoodsName(String goodsName) {
    this.goodsName = goodsName;
}
public double getGoodsPrice() {
    return goodsPrice;
}
public void setGoodsPrice(double goodsPrice) {
    this.goodsPrice = goodsPrice;
}
public String getGoodsType() {
    return goodsType;
}
public void setGoodsType(String goodsType) {
    this.goodsType = goodsType;
}
//添加商品
public void addGoods(){
Connection con = null;
    PreparedStatement ps = null;
    try {
        Class.forName("oracle.jdbc.driver.OracleDriver");
    } catch (ClassNotFoundException e) {
        e.printStackTrace();
    }
    try {
        con=DriverManager.getConnection("jdbc:oracle:thin:@localhost:1521:orcl",
                "system","system");
        ps=con.prepareStatement("insert into goodsInfo values(?,?,?,?)");
        ps.setInt(1, goodsId);                    //ps 调用 set 方法指定第 1 个通配符的值
        ps.setInt(1, goodsId);                    //ps 调用 set 方法指定第 2 个通配符的值
```

```
            ps.setInt(1, goodsId);                    //ps 调用 set 方法指定第 3 个通配符的值
            ps.setString(4, goodsType);               //ps 调用 set 方法指定第 4 个通配符的值
            ps.executeUpdate();
        }catch (SQLException e) {
            e.printStackTrace();
        }finally{
            try{
                if(ps!=null){
                    ps.close();
                }
                if(con!=null){
                    con.close();
                }
            }catch (SQLException e) {
                e.printStackTrace();
            }
        }
    }
    //获得所有商品信息
    public StringBuffer getQueryResult(){
        queryResult=new StringBuffer();
        Connection con = null;
        PreparedStatement ps = null;
        ResultSet rs=null;
        try {
            Class.forName("oracle.jdbc.driver.OracleDriver");
        } catch (ClassNotFoundException e) {
            e.printStackTrace();
        }
        try {
            con=DriverManager.getConnection("jdbc:oracle:thin:@localhost:1521:orcl",
                    "system","system");
            ps=con.prepareStatement("select * from goodsInfo");
            rs=ps.executeQuery();
            queryResult.append("<table border=1>");
                queryResult.append("<tr>");
                    queryResult.append("<th>goodsId</th>");
                    queryResult.append("<th>goodsName</th>");
                    queryResult.append("<th>goodsPrice</th>");
                    queryResult.append("<th>goodsType</th>");
                queryResult.append("</tr>");
            while(rs.next()){
                queryResult.append("<tr>");
                    queryResult.append("<td>"+rs.getString(1)+"</td>");
                    queryResult.append("<td>"+rs.getString(2)+"</td>");
```

```
                queryResult.append("<td>"+rs.getString(3)+"</td>");
                queryResult.append("<td>"+rs.getString(4)+"</td>");
            queryResult.append("</tr>");
        }
        queryResult.append("</table>");
    }catch (SQLException e) {
        e.printStackTrace();
    }finally{
        try{
            if(rs!=null){
                rs.close();
            }
            if(ps!=null){
                ps.close();
            }
            if(con!=null){
                con.close();
            }
        }catch (SQLException e) {
            e.printStackTrace();
        }
    }
    return queryResult;
  }
}
```

通过之前的学习，我们已经知道 Statement 在执行 executeQuery(String sql)、executeUpdate(String sql)等方法时，如果 SQL 语句有些部分是动态的数据，必须使用“+”连字符组成完整的 SQL 语句，十分不便。例如 6.3 节中的任务在添加商品时，必须按如下方式组成 SQL 语句：

```
String addSql="insert into goodsInfo values("+goodsId+",'"+goodsName+"',"+goodsPrice+
",'"+goodsType+"')";
st.executeUpdate(addSql);
```

PreparedStatement 对象被称为预处理语句对象，现在使用预处理语句不仅提高了数据库的访问效率，而且方便了程序的编写。预处理语句对象调用 executeUpdate()和 executeQuery()方法时不需要传递参数。例如：

```
int i=ps.executeUpdate();
```

或

```
ResultSet rs=ps.executeQuery();
```

6.5.3　拓展训练

编写两个 JSP 页面：inputPrepareQuery.jsp 和 showPrepareBy.jsp。用户可以在页面 inputPrepareQuery.jsp 中输入查询条件后，单击“查询”按钮。然后，在 showPrepareBy.jsp 页面中显示符合查询条件的商品信息。在本节任务的 bean(UsePrepare.java)中添加一个方法 getQueryPrepareResultBy()实现本题的条件查询功能(使用预处理语句实现查询)。页面运行效果如图 6-23 和图 6-24 所示。

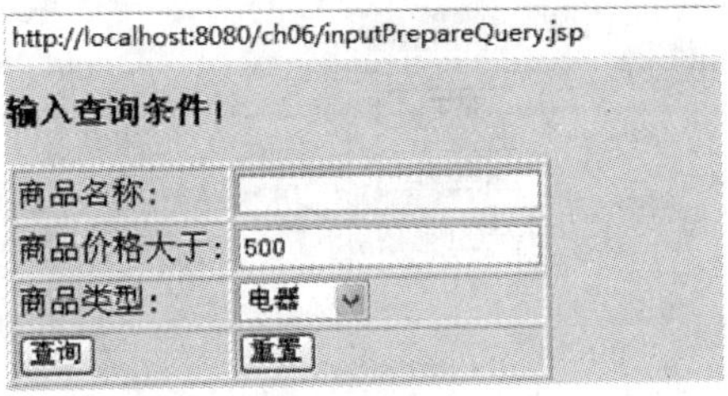
http://localhost:8080/ch06/inputPrepareQuery.jsp
输入查询条件：
商品名称：
商品价格大于：
500
商品类型：
电器
查询
重置
图 6-23　使用预处理条件查询

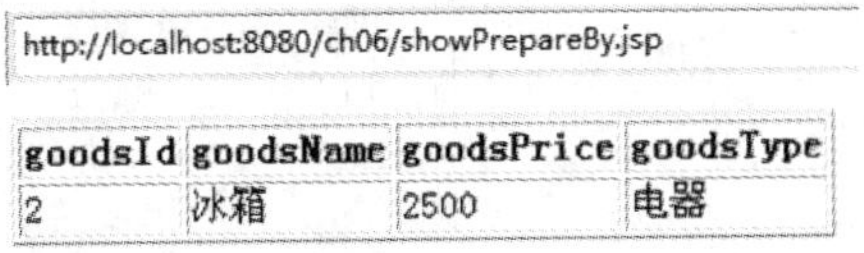
http://localhost:8080/ch06/showPrepareBy.jsp
goodsId
goodsName
goodsPrice
goodsType
2
冰箱
2500
电器
图 6-24　符合查询条件的记录

6.6　访问 Excel 电子表格

6.6.1　知识要点

当需要对 Excel 电子表格的内容进行增加、修改、查询和删除时，可以使用 JDBC-ODBC 桥接器的方式访问 Excel 电子表格。但访问 Excel 电子表格和访问关系数据库有所不同。访问 Excel 电子表格要经过以下两个步骤。

1. 创建 Excel 电子表格

使用 Microsoft Office Excel 2007 在物理硬盘“D:\”根目录下创建一个名为 bookInfo.xlsx 的电子表格，该 Excel 表格中有个名为 bookDetail 的工作表，具体内容如图 6-25 所示。

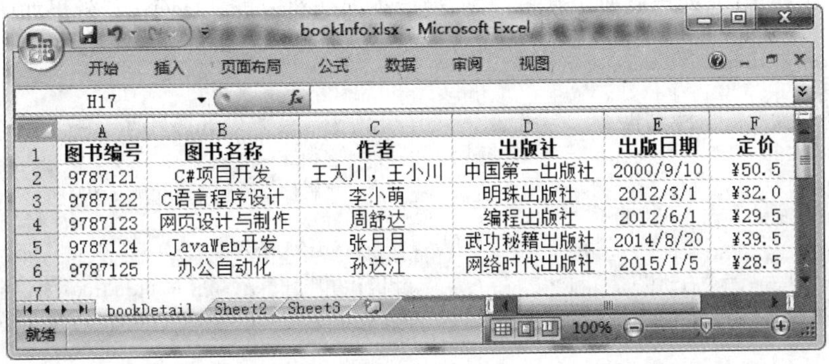
bookInfo.xlsx - Microsoft Excel
开始
插入
页面布局
公式
数据
审阅
视图
H17
图书编号 图书名称 作者 出版社 出版日期 定价
9787121 C#项目开发 王大川，王小川 中国第一出版社 2000/9/10 ¥50.5
9787122 C语言程序设计 李小萌 明珠出版社 2012/3/1 ¥32.0
9787123 网页设计与制作 周舒达 编程出版社 2012/6/1 ¥29.5
9787124 JavaWeb开发 张月月 武功秘籍出版社 2014/8/20 ¥39.5
9787125 办公自动化 孙达江 网络时代出版社 2015/1/5 ¥28.5
bookDetail
Sheet2
Sheet3
就绪
100%
图 6-25　student.xlsx 电子表格

2. 创建数据源

创建 Excel 电子表格之后，就可以按照 6.1 节中介绍的方法为它创建数据源了。首先为数据源选择的驱动程序是 Microsoft Excel Driver(*.xls、*.xlsx、*.xlsm、*.xlsb)，然后在如图 6-26 所示的对话框中将数据源命名为 bookSource，接下来单击“选择工作簿”按钮选择上一步中创建的“D:\bookInfo.xlsx”工作簿。还需要尤其注意的是，如果想通过 JDBC-ODBC 修改 Excel 电子表格，在设置数据源时，单击“选项”按钮，将“只读”属性设置为未选中状态。

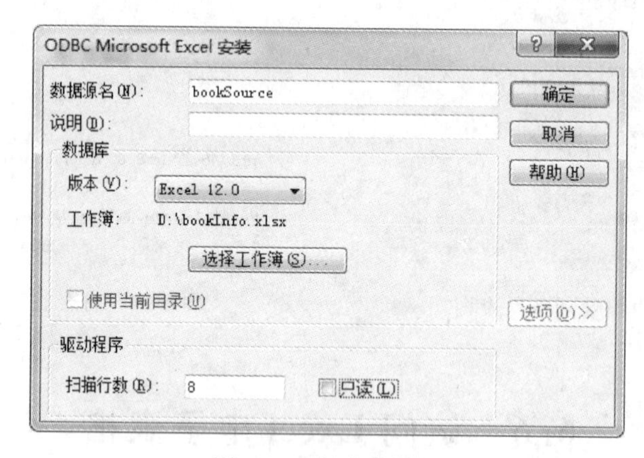

图 6-26 创建数据源

需要注意的是，访问电子表格时，把其中的工作表看成数据库中的表，如 bookInfo.xlsx 中的 bookDetail 工作表。使用 SQL 语句对工作表中的数据进行增加、删除、修改和查询时，要在表名前加“[”，在表名后加“$]”，如 select * from [bookDetail$]。

6.6.2 技能操作

使用 JDBC-ODBC 桥接器的方式访问 Excel 电子表格。具体任务如下：

编写一个 JSP 页面 readExcel.jsp，在该页面的 Java 程序片中首先增加一条记录到 studentScore 工作表中，然后修改了某条记录，最后查询全部记录。页面运行效果如图 6-27 所示。

http://localhost:8080/ch06/operateExcel.jsp

图书编号	图书名称	作者	出版社	出版日期	定价
9787121	C#项目开发	王大川，王小川	中国第一出版社	2000-09-10 00:00:00	50.5000
9787122	C语言程序设计	李小萌	中国第一出版社	2012-03-01 00:00:00	32.0000
9787123	网页设计与制作	周舒达	编程出版社	2012-06-01 00:00:00	29.5000
9787124	JavaWeb开发	张月月	武功秘籍出版社	2014-08-20 00:00:00	39.5000
9787125	办公自动化	孙达江	网络时代出版社	2015-01-05 00:00:00	28.5000
9787126	计算机网络	李三	启明星出版社	2015-04-01 00:00:00	30.0000

图 6-27 增加、修改和查询 Excel 表的记录

代码模板 operateExcel.jsp 如下：

```
<% @ page language = " java " contentType = " text/html; charset = GBK " pageEncoding =
"GBK"%>
<%@ page import="java.sql.*"%>
<html>
<head>
    <title>operateExcel.jsp</title>
</head>
<body bgcolor="lightblue">
    <%
        Connection con = null;
        Statement st = null;
        ResultSet rs = null;
        try {
            Class.forName("sun.jdbc.odbc.JdbcOdbcDriver");
        } catch (ClassNotFoundException e) {
            e.printStackTrace();
        }
        try {
            con = DriverManager.getConnection("jdbc:odbc:bookSource","", "");
            st=con.createStatement();
            //添加记录
            st.executeUpdate("insert into [bookDetail$]
            values('9787126','计算机网络','李三','启明星出版社','2015/4/1','¥30.0')");
            //修改记录
            st.executeUpdate("update [bookDetail$]
            set 出版社='中国第一出版社' where 图书编号='9787122'");
            //查询全部记录
            rs=st.executeQuery("select * from [bookDetail$]");
            out.print("<table border=1>");
            out.print("<tr>");
                out.print("<th>图书编号</th>");
                out.print("<th>图书名称</th>");
                out.print("<th>作者</th>");
                out.print("<th>出版社</th>");
                out.print("<th>出版日期</th>");
                out.print("<th>定价</th>");
            out.print("</tr>");
            while(rs.next()){
                out.print("<tr>");
                    out.print("<td>"+rs.getString(1)+"</td>");
                    out.print("<td>"+rs.getString(2)+"</td>");
                    out.print("<td>"+rs.getString(3)+"</td>");
                    out.print("<td>"+rs.getString(4)+"</td>");
                    out.print("<td>"+rs.getString(5)+"</td>");
```

```
                out.print("<td>"+rs.getString(6)+"</td>");
            out.print("</tr>");
        }
        out.print("</table>");
    } catch (SQLException e) {
        e.printStackTrace();
    }finally{
        try{
            if(rs!=null){
                rs.close();
            }
            if(st!=null){
                st.close();
            }
            if(con!=null){
                con.close();
            }
        }catch (SQLException e) {
            e.printStackTrace();
        }
    }
%>
</body>
</html>
```

说明：一个 Excel 电子表格可以有多个工作表，我们使用 JDBC-ODBC 可以访问该电子表格中的任何一个工作表，就像访问一个数据库中的任意一张表一样。

6.6.3 拓展训练

在 bookInfo.xlsx 电子表格中新建一个工作表 bookFare，详细内容如图 6-28 所示；编写一个 JSP 页面 practiceExcel.jsp，在该页面中显示 bookFare 表中的所有记录，如图 6-29 所示。

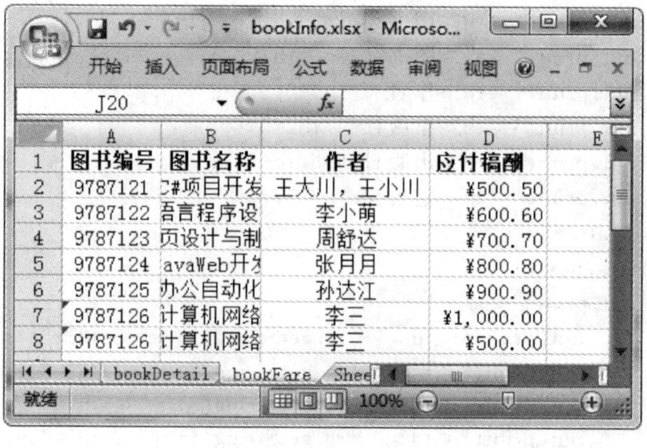

	A	B	C	D	E
1	图书编号	图书名称	作者	应付稿酬	
2	9787121	C#项目开发	王大川，王小川	¥500.50	
3	9787122	吾言程序设	李小萌	¥600.60	
4	9787123	页设计与制	周舒达	¥700.70	
5	9787124	avaWeb开	张月月	¥800.80	
6	9787125	办公自动化	孙达江	¥900.90	
7	9787126	计算机网络	李三	¥1,000.00	
8	9787126	计算机网络	李三	¥500.00	

图 6-28 工作表 bookFare

http://localhost:8080/ch06/bookFare.jsp

图书编号	图书名称	作者	应付稿酬
9787121	C#项目开发	王大川，王小川	500.5000
9787122	C语言程序设计	李小萌	600.6000
9787123	网页设计与制作	周舒达	700.7000
9787124	JavaWeb开发	张月月	800.8000
9787125	办公自动化	孙达江	900.9000
9787126	计算机网络	李三	1000.0000
9787126	计算机网络	李三	500.0000

图 6-29　practiceExcel.jsp 页面效果

6.7　使用数据库连接池

6.7.1　知识要点

和数据库建立连接是一个费时的活动，每次都要花费一定的时间，而且系统还要分配内存资源。这个时间对于一次或几次数据库操作，或许感觉不出系统有多大的开销。可是对于大型电子商务网站同时有几百人甚至几千人频繁地进行数据库连接操作的情况，势必占用很多的系统资源，网站的响应速度必定下降，严重时甚至会造成服务器的崩溃。因此，合理地建立数据库连接是非常重要的。

数据库连接池的基本思想是：为数据库连接建立一个“缓冲池”。预先在“缓冲池”中放入一定数量的连接，当需要建立数据库连接时，只需从“缓冲池”中取出一个，使用完毕之后再放回去。可以通过设定连接池最大连接数来防止系统无限度地与数据库连接。更为重要的是，通过连接池的管理机制监视数据库连接的数量及使用情况，为系统开发、测试和性能调整提供依据。其工作原理如图 6-30 所示。

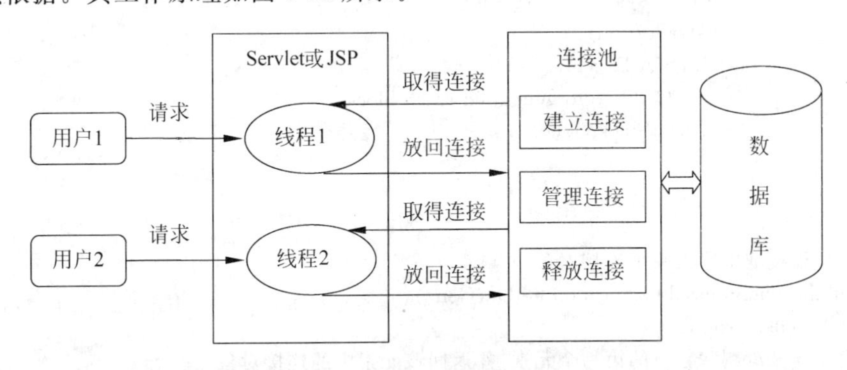

图 6-30　连接池的原理

6.7.2 技能操作

使用连接池连接数据库。具体任务如下：

编写一个 JSP 页面 conPool.jsp，在该页面中使用 scope 为 application 的 bean（由 ConnectionPool 类负责创建）。该 bean 创建时，将建立一定数量的连接对象。因此，所有的用户将共享这些连接对象。在 JSP 页面中使用 bean 获得一个连接对象，然后使用该连接对象访问数据库中的 goodsInfo 表（查询出商品价格大于 500 的商品）。页面运行效果如图 6-31 所示。

http://localhost:8080/ch06/conPool.jsp

商品编号	商品名称	商品价格	商品类别
2	冰箱	2500	电器
5	上衣	1800	服装

图 6-31　使用连接池连接数据库

代码模板 ConnectionPool.java 如下：

```
package db.connection.pool;
import java.sql.*;
import java.util.ArrayList;
public class ConnectionPool {
    //存放 Connection 对象的数组，数组被看成连接池
    ArrayList<Connection> list=new ArrayList<Connection>();
    //构造方法，创建 15 个连接对象，放到连接池中
    public ConnectionPool(){
        try {
            Class.forName("oracle.jdbc.driver.OracleDriver");
        } catch (ClassNotFoundException e) {
            e.printStackTrace();
        }
        for(int i=0;i<15;i++){
            try {
                Connection con=DriverManager.getConnection("jdbc:oracle:thin:@localhost:1521:
                orcl","system","system");
                list.add(con);
            } catch (SQLException e) {
                // TODO Auto-generated catch block
                e.printStackTrace();
            }
        }
    }
    //从连接池中取出一个连接对象
    public synchronized Connection getOneCon(){
        if(list.size()>0){
            //删除数组的第一个元素，并返回该元素中的连接对象
            return list.remove(0);
        }else{
```

```
            //连接对象被用完
            return null;
        }
    }
    //把连接对象放回连接池中
    public synchronized void releaseCon(Connection con){
        list.add(con);
    }
}
```

代码模板 conPool.jsp 如下：

```
<%@ page language="java" contentType="text/html; charset=GBK" pageEncoding=
"GBK"%>
<%@ page import="java.sql.*"%>
<%@ page import="db.connection.pool.*"%>
<html>
<head>
<title>使用连接池连接数据库</title>
</head>
<jsp:useBean id="conpool" class="db.connection.pool.ConnectionPool" scope="application"/>
<body bgcolor="AliceBlue">
<%
    Connection con=null;
    Statement st=null;
    ResultSet rs=null;
    try{
        //使用 conpool 对象调用 getOneCon 方法从连接池中获得一个连接对象
        con=conpool.getOneCon();
        if(con==null){
            out.print("人数过多,稍后访问");
            return;
        }
        st=con.createStatement();
        rs=st.executeQuery("select * from goodsInfo where goodsPrice>500");
        out.print("<table border=1>");
            out.print("<tr>");
                out.print("<th>商品编号</th>");
                out.print("<th>商品名称</th>");
                out.print("<th>商品价格</th>");
                out.print("<th>商品类别</th>");
            out.print("</tr>");
        while(rs.next()){
            out.print("<tr>");
                out.print("<td>"+rs.getString(1)+"</td>");
                out.print("<td>"+rs.getString(2)+"</td>");
```

```
                out.print("<td>"+rs.getString(3)+"</td>");
                out.print("<td>"+rs.getString(4)+"</td>");
            out.print("</tr>");
        }
        out.print("</table>");
    }catch(SQLException e){
        e.printStackTrace();
    }finally{
        try{
            if(rs!=null){
                rs.close();
            }
            if(st!=null){
                st.close();
            }
            if(con!=null){
            //使用 conpool 对象调用 releaseCon 方法把连接对象放回连接池中
                conpool.releaseCon(con);
            }
        }catch(SQLException e){
            e.printStackTrace();
        }
    }
%>
</body>
</html>
```

我们再打开一个新的浏览器窗口运行 conPool.jsp 页面时，会发现这一次访问的速度要比第一次快得多，而且也比前面介绍的访问 JSP 页面的速度要快，这是因为在第一次访问时连接池中没有可用连接，因此页面要等待创建一个新的连接，但是在第二次访问时连接池中就有一个可用连接了，可以直接使用这个连接来访问数据库。

6.7.3 拓展训练

编写一个 JSP 页面 praticePool.jsp，在该页面中使用和技能操作中同样的 bean 获得一个数据库连接对象，然后使用该连接对象查询 goodsInfo 表中的全部记录。

6.8 其他常见数据库的连接

6.8.1 知识要点

使用 JDBC-ODBC 桥接器的方式连接不同类型数据库的程序流程是类似的，只是在为 ODBC 数据源选择驱动程序时，选择对应的驱动程序即可，但有的数据库 ODBC 不支持，如

MySQL数据库。

使用纯Java数据库驱动程序连接不同类型数据库的程序流程和框架是基本相同的，需要重点关注的是，在连接各数据库时驱动程序加载部分的代码和连接部分的代码。下面分别介绍直接加载纯Java数据库驱动程序连接SQL Server 2005和MySQL 5.5。这里假定要访问的数据库名为mydatabase。

1. 连接SQL Server 2005

(1) 获取纯Java数据库驱动程序。

可以登录微软的官方网站 http://www.microsoft.com 下载 Microsoft SQL Server 2005 JDBC Driver 1.2，解压 Microsoft SQL Server 2005 jdbc driver1.2.exe 后得到 sqljdbc.jar 文件。然后，把 sqljdbc.jar 文件复制到 Web 应用程序的/WEB-INF/lib 文件夹中。

(2) 加载驱动程序。

```
Class.forName("com.microsoft.sqlserver.jdbc.SQLServerDriver");
```

(3) 建立连接。

```
Connection con=
        DriverManager. getConnection ( " jdbc: sqlserver://localhost: 1433; DatabaseName =
mydatabase ",
      "用户名", "密码");
```

2. 连接MySQL 5.5

(1) 获取纯Java数据库驱动程序。

可以登录MySQL的官方网站 http://www.mysql.com 下载驱动程序。这里下载的是 mysql-connector-java-5.0.6-bin.jar。然后，把 mysql-connector-java-5.0.6-bin.jar 文件复制到 Web 应用程序的/WEB-INF/lib 文件夹中。

(2) 加载驱动程序。

```
Class.forName("com.mysql.jdbc.Driver");
```

(3) 建立连接。

```
Connection con=
      DriverManager.getConnection("jdbc:mysql://localhost:3306/mydatabase","用户名","密码");
```

6.8.2 技能操作

使用纯Java数据库驱动程序连接其他常见数据库，如MySQL。具体任务如下：

- 安装MySQL 5.5。
- 创建数据库和数据表。
- 连接数据库并操作数据表。

1. 安装 MySQL 5.5

登录官网下载 MySQL 5.5 安装程序 mysql-5.5.19-win32.msi，按照默认安装即可。用户名和密码都设置为 root。

2. 创建数据库和数据表

MySQL 5.5 安装后，选择"开始"→"程序"→MySQL→MySQL Server 5.5→MySQL 5.5 Command Line Client 命令。在出现的界面中输入默认密码：root，成功启动 MySQL 监视器后，在 MS-DOS 窗口中出现"mysql>"字样，如图 6-32 所示。

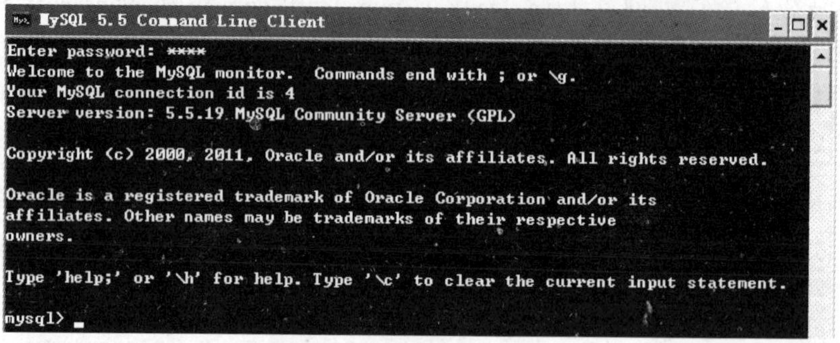

```
MySQL 5.5 Command Line Client
Enter password: ****
Welcome to the MySQL monitor.  Commands end with ; or \g.
Your MySQL connection id is 4
Server version: 5.5.19 MySQL Community Server (GPL)

Copyright (c) 2000, 2011, Oracle and/or its affiliates. All rights reserved.

Oracle is a registered trademark of Oracle Corporation and/or its
affiliates. Other names may be trademarks of their respective
owners.

Type 'help;' or '\h' for help. Type '\c' to clear the current input statement.

mysql> _
```

图 6-32　启动 MySQL 监视器

成功启动 MySQL 监视器后，就可以使用如下命令：

```
create database mydatabase;
```

创建数据库 mydatabase，为了在 mydatabase 数据库中创建表，必须先进入该数据库，命令如下：

```
use mydatabase
```

进入数据库 mydatabase 的操作如图 6-33 所示。

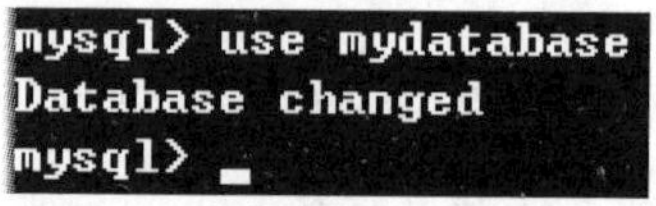

```
mysql> use mydatabase
Database changed
mysql> _
```

图 6-33　进入数据库

进入数据库后，就可以创建表 employee。创建表的命令如下：

```
create table employee (
    empNo varchar(10) not null,
    name varchar(30) not null,
    salary float not null,
    primary key(empNo)
);
insert into employee values('001','zhou',1200);
insert into employee values('002','wu',2500);
insert into employee values('003','zheng',2800 );
insert into employee values('004','wang',4800);
```

```
insert into employee values('005','zhao',1800);
insert into employee values('006','qian',7800);
commit;
```

3. 连接数据库并操作数据表

编写一个 JSP 页面 useMySQL.jsp，在该页面的 Java 程序片中连接 MySQL 数据库，并从 employee 表中查询出工资高于 2000 元的雇员。页面运行效果如图 6-34 所示。

http://localhost:8080/ch06/useMySQL.jsp

雇员编号	雇员姓名	工资
002	wu	2500
003	zheng	2800
004	wang	4800
006	qian	7800

图 6-34　工资高于 2000 元的雇员

代码模板 useMySQL.jsp 如下：

```
<%@ page language="java" contentType="text/html; charset=GBK"pageEncoding="GBK"%>
<%@ page import="java.sql.*"%>
<html>
<head>
<title>useMySQL.jsp</title>
</head>
<body bgcolor="lightblue">
    <%
        Connection con = null;
        Statement st = null;
        ResultSet rs = null;
        try {
            Class.forName("com.mysql.jdbc.Driver");    //加载 MySQL 的 Java 驱动
        } catch (ClassNotFoundException e) {
            e.printStackTrace();
        }
        try {
        //与 MySQL 建立连接,数据库名为 mydatabase,用户名和密码都为 root
        con=DriverManager.getConnection("jdbc:mysql://localhost:3306/mydatabase", "root", "root");
            st=con.createStatement();
            rs=st.executeQuery("select * from employee where salary>2000");
            out.print("<table border=1>");
            out.print("<tr>");
                out.print("<th>雇员编号</th>");
                out.print("<th>雇员姓名</th>");
```

```
                out.print("<th>工资</th>");
            out.print("</tr>");
            while(rs.next()){
                out.print("<tr>");
                    out.print("<td>"+rs.getString(1)+"</td>");
                    out.print("<td>"+rs.getString(2)+"</td>");
                    out.print("<td>"+rs.getString(3)+"</td>");
                out.print("</tr>");
            }
            out.print("</table>");
        } catch (SQLException e) {
            e.printStackTrace();
        }finally{
            try{
                if(rs!=null){
                    rs.close();
                }
                if(st!=null){
                    st.close();
                }
                if(con!=null){
                    con.close();
                }
            }catch (SQLException e) {
                e.printStackTrace();
            }
        }
    %>
</body>
</html>
```

6.8.3 拓展训练

1. 微软公司提供的连接 SQL Server 2005 的 JDBC 驱动程序是(　　)。

   A. com.mysql.jdbc.Driver

   B. sun.jdbc.odbc.JdbcOdbcDriver

   C. oracle.jdbc.driver.OracleDriver

   D. com.microsoft.sqlserver.jdbc.SQLServerDriver

2. 参考 6.8.2 节中的主要内容,使用 MySQL 创建一个数据库 yourdatabase,并在该数据库中创建一张表 student,然后编写程序操作该表。

6.9　小　　结

- JDBC 连接数据库有两种常用方式：建立 JDBC-ODBC 桥接器和加载纯 Java 驱动程序。
- 目前，有多种类型数据库。JDBC 不管连接哪种类型的数据库，连接方式都基本类似。需要重点关注的是在连接各数据库时驱动程序加载部分的代码和连接部分的代码。
- 数据库连接池的基本思想是：为数据库连接建立一个"缓冲池"。预先在"缓冲池"中放入一定数量的连接，当需要建立数据库连接时，只需从"缓冲池"中取出一个，使用完毕之后再放回去。可以通过设定连接池最大连接数来防止系统无限度地与数据库连接。更为重要的是，通过连接池的管理机制监视数据库连接的数量及使用情况，为系统开发、测试和性能调整提供依据。
- 预处理语句对象不仅大大提高了数据库的执行速度，而且方便编写程序。

第7章

JSP与Servlet

Java Servlet 的核心思想就是在 Web 服务器端创建用来响应用户请求的对象，该对象被称为一个 Servlet 对象。JSP 技术以 Java Servlet 为基础，当客户请求一个 JSP 页面时，Web 服务器(如 Tomcat 服务器)会自动生成一个对应的 Java 文件，编译该 Java 文件，并用编译得到的字节码文件在服务器端创建一个 Servlet 对象。但大多数 Web 应用需要 Servlet 对象具有特定的功能，这时就需要 Web 开发人员自己编写创建 Servlet 对象的类。本章将重点介绍如何编写 Servlet 类与如何使用 Servlet 类。

7.1 编写 Servlet

7.1.1 知识要点

编写一个 Servlet 类很简单，只要继承 javax.servlet.http 包中的 HttpServlet 类，并重写响应 HTTP 请求的方法即可。HttpServlet 类实现了 Servlet 接口，实现了响应用户的接口方法。HttpServlet 类的一个子类习惯上称为一个 Servlet 类，这样的子类创建的对象习惯上称为 servlet 对象。

7.1.2 技能操作

编写一个简单的 Servlet 类 MyFirstServlet，用户请求这个 servlet 对象时，就会在浏览器中看到“人生第一个 Servlet 类”这样的响应信息。

代码模板 MyFirstServlet.java 如下：

```
package servlet;
import java.io.*;
import javax.servlet.*;
import javax.servlet.http.*;
public class MyFirstServlet extends HttpServlet {
    public void init(ServletConfig config) throws ServletException{
        super.init(config);
    }
    public void service(HttpServletRequest request, HttpServletResponse response)
            throws IOException{
```

```
            //设置响应的内容类型
            response.setContentType("text/html;charset=utf-8");
            //取得输出对象
            PrintWriter out=response.getWriter();
            out.println("<html><body>");
            //在浏览器中显示：人生第一个Servlet类
            out.println("人生第一个Servlet类");
            out.println("</body></html>");
        }
    }
```

编写Servlet类时必须有包名。也就是说，必须在包中编写Servlet类。在本章中我们新建一个Web工程ch07，所有的Servlet类放在src下servlet包中。

技能操作中Servlet类的源文件MyFirstServlet.java保存在Eclipse的Web工程ch07的src下servlet包里面。MyFirstServlet.java源文件由Eclipse自动编译生成字节码文件MyFirstServlet.class，保存在build\classes\servlet里面。Servlet类的源文件与字节码文件保存目录如图7-1所示。编写完Servlet类的源文件，并编译了源文件，这时是不是就可以运行servlet对象呢？不可以，需要部署servlet之后才可以运行servlet对象。

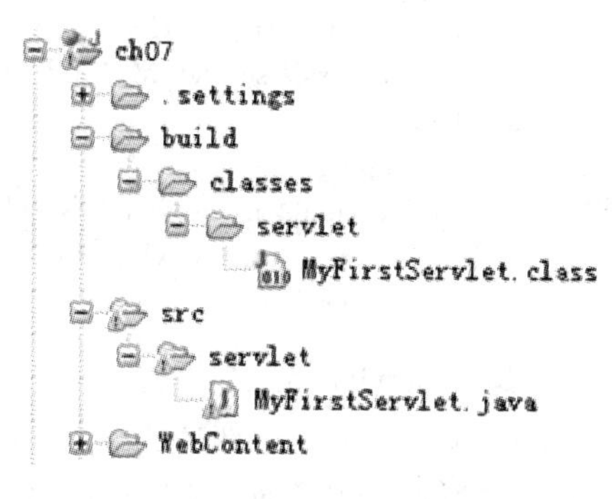

图7-1 保存目录

7.1.3 拓展训练

1. servlet对象是在服务器端被创建的还是在用户端被创建的？

2. 编写一个简单的Servlet类YourFirstServlet。用户请求这个servlet对象时，就会在浏览器中看到“人生第一个Servlet类”这样的响应信息，并在Web工程中找到该Servlet类对应的字节码文件。

7.2 部署与运行Servlet

7.2.1 知识要点

要想让Web服务器用Servlet类编译后的字节码文件创建servlet对象的话，必须为Web服务器编写一个部署文件。这个部署文件是一个XML文件，名字是web.xml，该文件由Web服务器负责管理。在ch07\ WebContent\WEB-INF目录里找到web.xml文件，可以在该文件里部署自己的servlet。

7.2.2 技能操作

掌握部署与运行 servlet 的方法，具体任务如下：

- 部署 servlet。
- 运行 servlet。

1. 部署 servlet

为了在 web. xml 文件里部署 7. 1 节中的 MyFirstServlet，需要在 web. xml 文件里找到<web-app></web-app>标记，然后在<web-app></web-app>标记中添加如下内容：

```
<servlet>
    <servlet-name>myServlet</servlet-name>
    <servlet-class>servlet. MyFirstServlet</servlet-class>
</servlet>
<servlet-mapping>
    <servlet-name>myServlet</servlet-name>
    <url-pattern>/firstServlet</url-pattern>
</servlet-mapping>
```

2. 运行 servlet

在 web. xml 文件中部署完 servlet 之后，就可以运行 servlet 了。servlet 第一次被直接访问(可以通过 JSP 页面间接访问)时，需要把它发布到 Web 服务器上(选中 Servlet 类的源文件，右击，选择 Run As→Run on Server 命令)。这时，在 Eclipse 内嵌的浏览器中看到如图 7-2 所示的画面。

http://localhost:8080/ch07/firstServlet

人生第一个Servlet类

图 7-2　servlet 运行效果

把 servlet 发布到 Web 服务器上之后，也可以在 IE 浏览器的地址栏中输入：

```
http://localhost:8080/ch07/firstServlet
```

来请求运行 servlet。

为了更好地理解本例，将涉及到的相关知识说明如下：

(1) web. xml 文件中的具体内容及其作用。

① 根标记<web-app>。

XML 文件中必须有一个根标记，web. xml 的根标记是<web-app>。

② <servlet>标记及其子标记。

web. xml 文件中可以有若干个<servlet>标记，该标记的内容由 Web 服务器负责处理。<servlet>标记中有两个子标记：<servlet-name>和<servlet-class>，其中<servlet-name>子标记的内容是 Web 服务器创建的 servlet 对象的名字。web. xml 文件中可以有若干个<servlet>标记，但要求其<servlet-name>子标记的内容互不相同。

<servlet-class>子标记的内容指定Web服务器用哪个类来创建servlet对象，如果servlet对象已经创建，那么Web服务器就不再使用指定的类创建。

③ <servlet-mapping>标记及其子标记。

web.xml文件中若出现一个<servlet>标记就会对应地出现一个<servlet-mapping>标记。<servlet-mapping>标记中也有两个子标记：<servlet-name>和<url-pattern>。其中<servlet-name>子标记的内容是Web服务器创建的servlet对象的名字（该名字必须和<servlet>标记的子标记<servlet-name>的内容相同）；<url-pattern>子标记用来指定用户用怎样的模式请求servlet对象。例如，<url-pattern>子标记的内容是/firstServlet，用户要请求服务器运行servlet对象myServlet为其服务，那么在IE浏览器的地址栏中输入：

```
http://localhost:8080/ch07/firstServlet
```

一个Web服务的web.xml文件负责管理该Web服务的servlet对象，当Web服务需要提供更多的servlet对象时，只要在web.xml文件中添加<servlet>和<servlet-mapping>标记就可以。

（2）servlet的生命周期。

一个servlet对象的生命周期主要由下列三个过程组成。

① 初始化servlet对象。

当servlet对象第一次被请求加载时，服务器会创建一个servlet对象，该servlet对象调用init方法完成必要的初始化工作。

② service方法响应请求。

创建的servlet对象再调用service方法响应客户的请求。

③ servlet对象死亡。

当服务器关闭时，servlet对象调用destroy方法使自己消亡。

从上面三个过程来看，init方法只能被调用一次，即在servlet第一次被请求加载时调用该方法。当客户请求servlet服务时，服务器将启动一个新的线程，在该线程中，servlet对象调用service方法响应客户的请求。那么多个客户请求servlet服务时，服务器会怎么办呢？服务器会为每个客户启动一个新的线程，在每个线程中，servlet对象调用service方法响应客户的请求。也就是说，每个客户请求都会导致service方法被调用执行，分别运行在不同的线程中。

（3）Servlet类中的方法。

① init方法。

init方法是HttpServlet类中的方法，可以在Servlet类中被重写。init方法的声明格式如下：

```
public void init(ServletConfig config) throws ServletException
```

servlet 对象第一次被请求加载时，服务器创建一个 servlet 对象，该对象调用 init 方法完成必要的初始化工作。

② service 方法。

service 方法是 HttpServlet 类中的方法，可以在 Servlet 类中被重写。service 方法的声明格式如下：

```
public void service(HttpServletRequest request, HttpServletResponse response)throws IOException
```

当 servlet 对象成功创建后，servlet 对象就调用 service 方法来处理用户的请求并返回响应。与 init 方法不同的是，init 方法只被调用一次，而 service 方法可能被多次调用。我们知道，当客户请求 servlet 对象时，服务器将启动一个新的线程，在该线程中，servlet 对象调用 service 方法响应客户的请求。换句话说，每个客户的每次请求都导致 service 方法被调用执行，调用过程运行在不同的线程中，互不干扰。

③ destroy 方法。

destroy 方法是 HttpServlet 类中的方法。一个 Servlet 类可直接继承该方法，一般不需要重写。destroy 方法的声明格式如下：

```
public void destroy()
```

当服务器终止服务时，destroy 方法会被执行，使 servlet 对象消亡。

7.2.3 拓展训练

1. 简述 servlet 的生命周期与运行原理。

2. servlet 对象初始化时是调用 init 方法还是 service 方法？

3. 编写一个简单的 Servlet 类 Pratice2Servlet，在 web.xml 中部署该 servlet 并运行它。当用户通过在 IE 浏览器地址栏中输入 http://localhost:8080/ch07/pratice2 请求这个 servlet 对象时，就会在浏览器中看到“部署与运行 servlet”这样的响应信息。

7.3 通过 JSP 页面访问 servlet

可以通过 JSP 页面的表单或超链接请求某个 servlet。通过 JSP 页面访问 servlet 的好处是，JSP 页面负责页面的静态信息处理，动态信息处理由 servlet 完成。本章所涉及的 JSP 页面均保存在 ch07\WebContent 目录里。

7.3.1 知识要点

1. 通过表单访问 servlet

假设在 JSP 页面中，有如下表单：

```
<form action="isLogin" method="post">
    ...
</form>
```

那么该表单的处理程序(action)就是一个 servlet,为该 servlet 部署时,web. xml 文件中的标记<servlet-mapping>的子标记<url-pattern>的内容是"/isLogin"。

2. 通过超链接访问 servlet

在 JSP 页面中,可以单击超链接,访问 servlet 对象,同时也可以通过超链接向 servlet 提交信息,例如,<a href="isLogin? user=tangtang&&pwd=123456">查看用户名和密码</a>,"查看用户名和密码"这个超链接就把 user=tangtang 和 pwd=123456 两个信息提交给 servlet 处理。

7.3.2 技能操作

灵活使用 JSP 页面访问 servlet 对象,具体任务如下:

编写一个 JSP 页面 login. jsp,在该页面中通过表单向名字为 login 的 servlet 对象(由 LoginServlet 类负责创建)提交用户名和密码,servlet 负责判断输入的用户名和密码是否正确,并把判断结果返回给客户。需要为 web. xml 文件添加如下的子标记:

```
<servlet>
    <servlet-name>login</servlet-name>
    <servlet-class>servlet. LoginServlet</servlet-class>
</servlet>
<servlet-mapping>
    <servlet-name>login</servlet-name>
    <url-pattern>/isLogin</url-pattern>
</servlet-mapping>
```

页面运行效果如图 7-3、图 7-4 和图 7-5 所示。

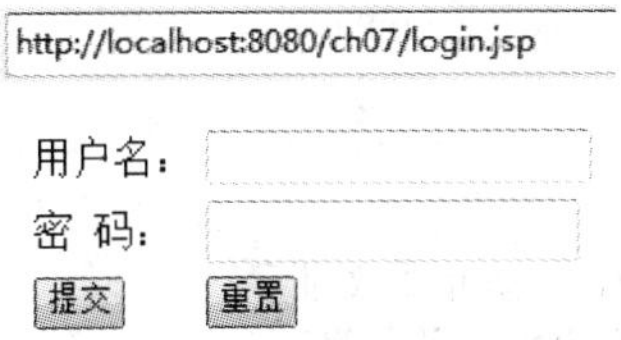

http://localhost:8080/ch07/login.jsp

用户名:

密 码:

提交　重置

图 7-3　信息输入页面

http://localhost:8080/ch07/isLogin

信息输入错误

图 7-4　servlet 错误信息

http://localhost:8080/ch07/isLogin

信息输入正确

图 7-5　servlet 正确信息

代码模板 login. jsp 如下:

```
<%@ page language = "java" contentType = "text/html; charset = GBK" pageEncoding =
"GBK"%>
<html>
    <head>
```

```
        <title>login.jsp</title>
    </head>
    <body>
        <form action="isLogin" method="post">
        <table>
            <tr>
                <td>用户名：</td>
                <td><input type="text" name="user"/></td>
            </tr>
            <tr>
                <td>密码：</td>
                <td><input type="password" name="pwd"/></td>
            </tr>
            <tr>
                <td><input type="submit" value="提交"/></td>
                <td><input type="reset" value="重置"/></td>
            </tr>
        </table>
    </form>
    </body>
    </html>
```

代码模板 LoginServlet.java 如下：

```
package servlet;
import java.io.*;
import javax.servlet.*;
import javax.servlet.http.*;
public class LoginServlet extends HttpServlet {
    public void init(ServletConfig config) throws ServletException{
        super.init(config);
    }
    public void service(HttpServletRequest request, HttpServletResponse response)
            throws IOException{
        response.setContentType("text/html;charset=GBK");
        PrintWriter out=response.getWriter();
        request.setCharacterEncoding("GBK");            //设置编码,防止中文乱码
        String name=request.getParameter("user");      //获取客户提交的信息
        String password=request.getParameter("pwd"); //获取客户提交的信息
        out.println("<html><body>");
        if(name==null||name.length()==0){
            out.println("请输入用户名");
        }
        else if(password==null||password.length()==0){
            out.println("请输入密码");
        }
```

```
        else if(name.length()>0&&password.length()>0){
            if(name.equals("tangtang")&&password.equals("123456")){
                out.println("信息输入正确");
            }else{
                out.println("信息输入错误");
            }
        }
        out.println("</body></html>");
    }
}
```

需要特别注意的是，如果 servlet 的请求格式是“/XXX”（请求格式就是 web.xml 文件中的标记<servlet-mapping>的子标记<url-pattern>的内容），那么 JSP 页面请求 servlet 时，必须写成“XXX”，不可以写成“/XXX”，否则将变成请求服务器（Tomcat）root 目录下的某个 servlet。

7.3.3　拓展训练

把技能操作任务中 login.jsp 的信息提交方式（表单提交）改成超链接提交方式来访问 servlet。

提示：超链接访问 servlet 格式如下：

```
<a href="isLogin?user=tangtang&&pwd=123456">查看用户名和密码</a>
```

7.4　doGet 和 doPost 方法

在编写 Servlet 类时，经常会重写 HttpServlet 类中的 doGet 和 doPost 方法，用来处理客户的请求并作出响应。

7.4.1　知识要点

当服务器接收到一个 servlet 请求时，就会产生一个新线程，在这个线程中让 servlet 对象调用 service 方法为请求作出响应。service 方法首先检查 HTTP 请求类型（get 或 post），并在 service 方法中根据用户的请求方式，相应地再调用 doGet 或 doPost 方法。

HTTP 请求类型为 get 方式时，service 方法调用 doGet 方法响应用户请求；HTTP 请求类型为 post 方式时，service 方法调用 doPost 方法响应用户请求。因此，在 Servlet 类中，没有必要重写 service 方法，直接继承即可。

在 Servlet 类中重写 doGet 或 doPost 方法来响应用户的请求，这样可以增加响应的灵活性，减轻服务器的负担。

7.4.2 技能操作

掌握 doGet 和 doPost 方法的调用原理，具体任务如下：

编写一个 JSP 页面 input.jsp，在该页面中使用表单向 servlet 对象 computer 提交矩形的长与宽的值。computer(由 GetLengthOrAreaServlet 负责创建)处理手段依赖表单提交数据的方式，当提交方式为 get 时，computer 对象计算矩形的周长；当提交方式为 post 时，computer 对象计算矩形的面积。需要为 web.xml 文件添加如下的子标记：

```
<servlet>
    <servlet-name>computer</servlet-name>
    <servlet-class>servlet.GetLengthOrAreaServlet</servlet-class>
</servlet>
<servlet-mapping>
    <servlet-name>computer</servlet-name>
    <url-pattern>/showLengthOrArea</url-pattern>
</servlet-mapping>
```

页面运行效果如图 7-6 至图 7-8 所示。

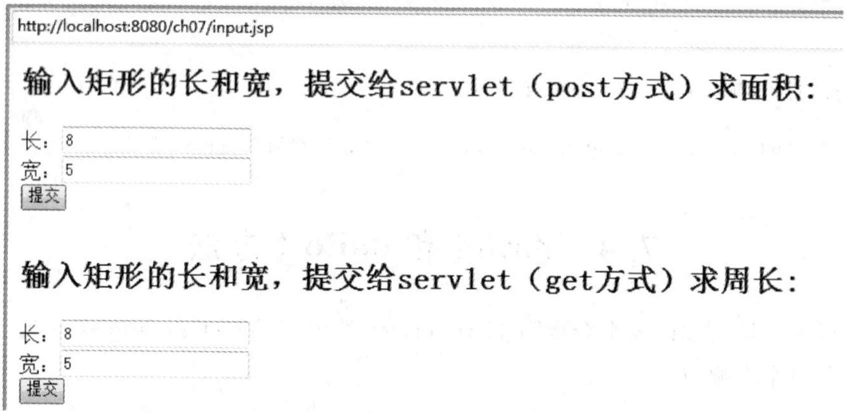

http://localhost:8080/ch07/input.jsp

输入矩形的长和宽，提交给servlet（post方式）求面积：

长：8
宽：5
提交

输入矩形的长和宽，提交给servlet（get方式）求周长：

长：8
宽：5
提交

图 7-6　信息输入页面

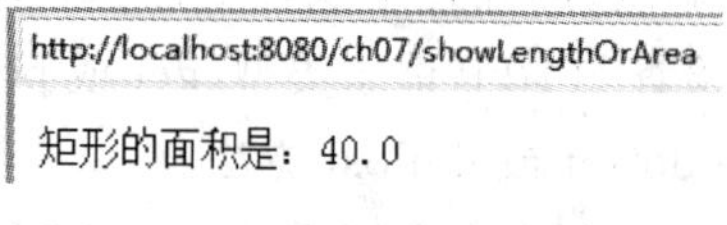

http://localhost:8080/ch07/showLengthOrArea

矩形的面积是：40.0

图 7-7　post 方式提交获得矩形面积

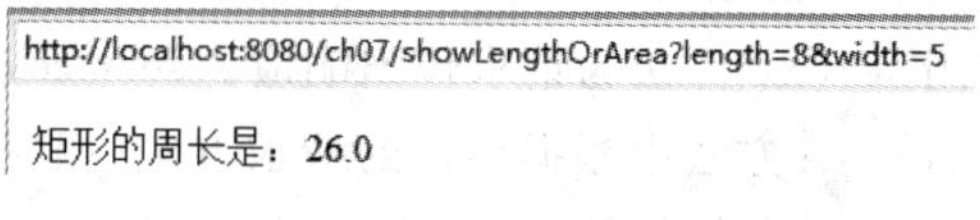

http://localhost:8080/ch07/showLengthOrArea?length=8&width=5

矩形的周长是：26.0

图 7-8　get 方式提交获得矩形周长

代码模板 input.jsp 如下：

```
<%@ page language="java" contentType="text/html; charset=GBK" pageEncoding=
"GBK"%>
```

```
<html>
<head>
  <title>input.jsp</title>
</head>
<body>
  <h2>输入矩形的长和宽,提交给 servlet(post 方式)求面积:</h2>
  <form action="showLengthOrArea" method="post">
    长: <input type="text" name="length"/><br/>
    宽: <input type="text" name="width"/><br/>
  <input type="submit" value="提交"/>
</form>
  <br/>
<h2>输入矩形的长和宽,提交给 servlet(get 方式)求周长:</h2>
<form action="showLengthOrArea" method="get">
    长: <input type="text" name="length"/><br/>
    宽: <input type="text" name="width"/><br/>
    <input type="submit" value="提交"/>
    </form>
  </body>
</html>
```

代码模板 GetLengthOrAreaServlet.java 如下:

```
package servlet;
import java.io.*;
import javax.servlet.*;
import javax.servlet.http.*;
public class GetLengthOrAreaServlet extends HttpServlet {
    public void init(ServletConfig config) throws ServletException{
        super.init(config);
    }
    public void doPost(HttpServletRequest request, HttpServletResponse
            response) throws ServletException, IOException{
        response.setContentType("text/html;charset=GBK");
        PrintWriter out=response.getWriter();
        String l=request.getParameter("length");
        String w=request.getParameter("width");
        out.println("<html><body>");
        double m=0,n=0;
        try{
            m=Double.parseDouble(l);
            n=Double.parseDouble(w);
            out.println("矩形的面积是: "+m*n);      //计算矩形的面积,并在浏览器上输出
        }catch(NumberFormatException e){
            out.println("请输入数字字符!");
        }
```

```
        out.println("</body></html>");
    }
    public void doGet(HttpServletRequest request, HttpServletResponse
            response) throws ServletException, IOException{
        response.setContentType("text/html;charset=utf-8");
        PrintWriter out=response.getWriter();
        String l=request.getParameter("length");
        String w=request.getParameter("width");
        out.println("<html><body>");
        double m=0,n=0;
        try{
            m=Double.parseDouble(l);
            n=Double.parseDouble(w);
            out.println("矩形的周长是："+(m+n)*2); //计算矩形的周长,并在浏览器上输出
        }catch(NumberFormatException e){
            out.println("请输入数字字符!");
        }
        out.println("</body></html>");
    }
}
```

一般情况下，不论用户请求类型是 get 还是 post，服务器的处理过程完全相同，那么可以只在 doPost 方法中编写处理过程，而在 doGet 方法中再调用 doPost 方法；或只在 doGet 方法中编写处理过程，而在 doPost 方法中再调用 doGet 方法。

7.4.3 拓展训练

编写一个 JSP 页面 pratice7_4.jsp，在该 JSP 页面中用户可以使用表单向 servlet 对象 pratice 提交矩形的长与宽的值。pratice(由 PraticeServlet 类负责创建)处理数据的手段不依赖表单提交数据的方式，即不论 post 还是 get，处理数据的手段相同，都是计算矩形的周长。

7.5 重定向与转发

重定向是将用户从当前 JSP 页面或 servlet 定向到另一个 JSP 页面或 servlet，以前的 request 中存放的信息全部失效，并进入一个新的 request 作用域；而转发则是将用户对当前 JSP 页面或 servlet 的请求转发给另一个 JSP 页面或 servlet，以前的 request 中存放的信息不会失效。

7.5.1 知识要点

1. 重定向

重定向方法 public void sendRedirect(String location)是 HttpServletResponse 类中的

方法。重定向的目标页面或 servlet(由参数 location 指定)无法从以前的 request 对象获取用户提交的数据。

2. 转发

javax.servlet.RequestDispatcher 对象可以把用户对当前 JSP 页面或 servlet 的请求转发给另一个 JSP 页面或 servlet。实现转发需要两个步骤:

- 获得 RequestDispatcher 对象。

在当前 JSP 页面或 servlet 中,使用 request 对象调用

```
public RequestDispatcher getRequestDispatcher(String url)
```

方法返回一个 RequestDispatcher 对象,其中参数 url 就是要转发的 JSP 页面或 servlet 的地址,例如:

```
RequestDispatcher dis=request.getRequestDispatcher("dologin");
```

- RequestDispatcher 对象调用 forward 方法实现转发。

获得 RequestDispatcher 对象之后,就可以使用该对象调用

```
public void forward(ServletRequest request, ServletResponse response)
```

方法将用户对当前 JSP 页面或 servlet 的请求转发给 RequestDispatcher 对象所指定的 JSP 页面或 servlet。例如:

```
dis.forward(request,response);
```

将用户对当前 JSP 页面或 servlet 的请求转变成对 dologin(servlet)的请求。

7.5.2　技能操作

理解重定向与转发的区别,掌握重定向与转发的实现方法,具体任务如下:

编写 JSP 页面 redirectForward.jsp,在该 JSP 页面中通过表单向名为 rforward 的 servlet 对象(由 RedirectForwardServlet 类负责创建)提交用户名和密码。如果用户输入的数据不完整(没有输入用户名或密码),那么 rforward 就将用户重定向到 redirectForward.jsp 页面;如果用户输入的数据完整,rforward 就将用户对 redirectForward.jsp 页面的请求转发给名字为 show 的 servlet 对象(由 ShowServlet 类负责创建),该 servlet 对象显示用户输入的信息。需要为 web.xml 文件添加如下的子标记:

```
<servlet>
    <servlet-name>rforward</servlet-name>
    <servlet-class>servlet.RedirectForwardServlet</servlet-class>
</servlet>
<servlet>
    <servlet-name>show</servlet-name>
```

```
    <servlet-class>servlet.ShowServlet</servlet-class>
</servlet>
<servlet-mapping>
    <servlet-name>rforward</servlet-name>
    <url-pattern>/rfLogin</url-pattern>
</servlet-mapping>
<servlet-mapping>
    <servlet-name>show</servlet-name>
    <url-pattern>/yourMessage</url-pattern>
</servlet-mapping>
```

代码模板 redirectForward.jsp 如下：

```
<%@ page language="java" contentType="text/html; charset=GBK" pageEncoding=
"GBK"%>
<html>
    <head>
        <title>redirectForward.jsp</title>
    </head>
    <body>
        <form action="rfLogin" method="post">
        <table>
            <tr>
                <td>用户名: </td>
                <td><input type="text" name="user"/></td>
            </tr>
            <tr>
                <td>密码: </td>
                <td><input type="password" name="pwd"/></td>
            </tr>
            <tr>
                <td><input type="submit" value="提交"/></td>
                <td><input type="reset" value="重置"/></td>
            </tr>
        </table>
      </form>
    </body>
</html>
```

代码模板 RedirectForwardServlet.java 如下：

```
package servlet;
import java.io.IOException;
import javax.servlet.RequestDispatcher;
import javax.servlet.ServletConfig;
```

```
import javax.servlet.ServletException;
import javax.servlet.http.HttpServlet;
import javax.servlet.http.HttpServletRequest;
import javax.servlet.http.HttpServletResponse;
public class RedirectForwardServlet extends HttpServlet {
    public void init(ServletConfig config) throws ServletException{
        super.init(config);
    }
    public void doPost(HttpServletRequest request, HttpServletResponse
            response) throws ServletException, IOException{
        String name=request.getParameter("user");
        String password=request.getParameter("pwd");
        if(name==null||name.length()==0){
            response.sendRedirect("redirectForward.jsp");
            //使用 response 调用 sendRedirect 方法重定向到 redirectForward.jsp
        }
        else if(password==null||password.length()==0){
            response.sendRedirect("redirectForward.jsp");
            //使用 response 调用 sendRedirect 方法重定向到 redirectForward.jsp
        }
        else if(name.length()>0&&password.length()>0){
            RequestDispatcher dis=request.getRequestDispatcher("yourMessage");
            //获得 RequestDispatcher 对象 dis,转发到 servlet 对象 yourMessage
            dis.forward(request, response);
            //dis 对象调用 forward 方法实现转发
        }
    }
    public void doGet(HttpServletRequest request, HttpServletResponse
            response) throws ServletException, IOException{
        doPost(request, response);
    }
}
```

代码模板 ShowServlet.java 如下：

```
package servlet;
import java.io.*;
import javax.servlet.*;
import javax.servlet.http.*;
public class ShowServlet extends HttpServlet {
    public void init(ServletConfig config) throws ServletException{
        super.init(config);
    }
    public void doPost(HttpServletRequest request, HttpServletResponse
            response) throws ServletException, IOException{
        response.setContentType("text/html;charset=GBK");
```

```
            PrintWriter out=response.getWriter();
            String name=request.getParameter("user");
            String password=request.getParameter("pwd");
            byte b[]=name.getBytes("ISO-8859-1");
            name=new String(b,"UTF-8");
            byte b1[]=password.getBytes("ISO-8859-1");
            password=new String(b1,"UTF-8");
            out.println("您的用户名是: "+name);
            out.println("<br>您的密码是: "+password);
        }
        public void doGet(HttpServletRequest request,HttpServletResponse
                response) throws ServletException,IOException{
            doPost(request,response);
        }
    }
```

和重定向方法不同的是,使用转发时,用户在浏览器的地址栏中不能看到 forward 方法所转发的页面地址或 servlet 的地址,只能看到它们的运行效果。用户在浏览器的地址栏中所看到的仍然是当前 JSP 页面或 servlet 的地址。

7.5.3 拓展训练

1. 什么是转发?什么是重定向?它们有什么区别?

2. 试着把技能操作任务中的转发改成重定向,然后运行 redirectForward.jsp 页面,看看运行结果是什么样的,为什么是这样的结果?

7.6 session 会话管理

7.6.1 知识要点

在 servlet 中可以使用 request 对象来调用 getSession 方法获取用户的会话对象 session,例如:

```
HttpSession session=request.getSession(true);
```

一个用户在同一个 Web 服务中的不同 servlet 对象中获取的 session 对象是完全相同的,但是不同用户的 session 对象互不相同。

7.6.2 技能操作

掌握如何在 servlet 中获取用户的会话对象 session,具体任务如下:

编写一个 JSP 页面 useSession.jsp,在该页面中通过表单向名字为 useSession 的 servlet 对象(由 UseSessionServlet 类负责创建)提交用户名,useSession 将用户名存入用户的 session 对象中，然后用户请求另一个 servlet 对象 showName(由 ShowNameServlet 类负责创建)，showName 从用户的 session 对象中取出存储的用户名，并显示在浏览器中。需要为 web.xml 文件添加如下的子标记：

```
<servlet>
    <servlet-name>useSession</servlet-name>
    <servlet-class>servlet.UseSessionServlet</servlet-class>
  </servlet>
<servlet>
    <servlet-name>showName</servlet-name>
    <servlet-class>servlet.ShowNameServlet</servlet-class>
</servlet>
<servlet-mapping>
    <servlet-name>useSession</servlet-name>
    <url-pattern>/sendMyName</url-pattern>
</servlet-mapping>
<servlet-mapping>
    <servlet-name>showName</servlet-name>
    <url-pattern>/showMyName</url-pattern>
</servlet-mapping>
```

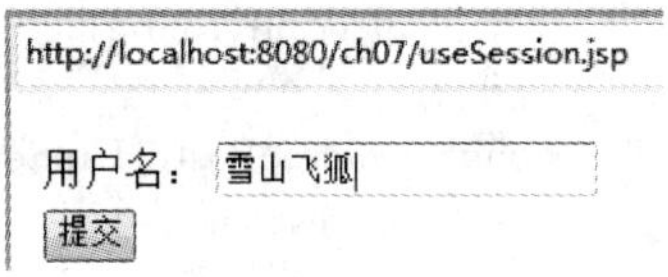

程序运行效果如图 7-9 至图 7-11 所示。

http://localhost:8080/ch07/useSession.jsp

用户名：雪山飞狐

提交

图 7-9　信息输入页面

http://localhost:8080/ch07/sendMyName

您请求的servlet对象是：useSession
您的会话ID是：F97E2DB26869F34B86D502BDAB7FD41D
请单击请求另一个servlet：
请求另一个servlet

图 7-10　获取会话并存储数据

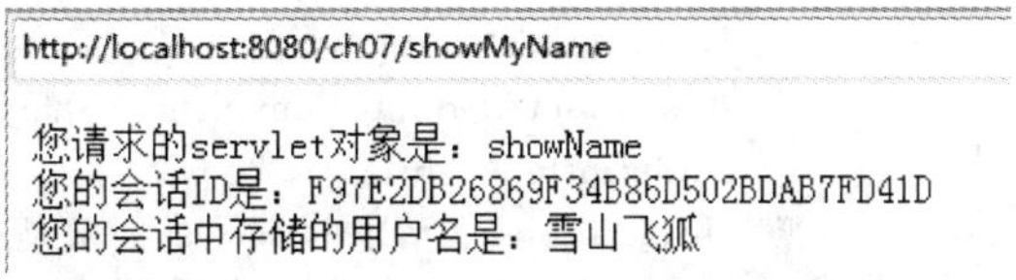

http://localhost:8080/ch07/showMyName

您请求的servlet对象是：showName
您的会话ID是：F97E2DB26869F34B86D502BDAB7FD41D
您的会话中存储的用户名是：雪山飞狐

图 7-11　获取会话中的数据并显示

代码模板 useSession.jsp 如下：

```
<%@ page language="java" contentType="text/html; charset=GBK" pageEncoding=
"GBK"%>
<html>
    <head>
        <title>useSession.jsp</title>
    </head>
    <body>
        <form action="sendMyName" method="post">
            <table>
                <tr>
                    <td>用户名：</td>
                    <td><input type="text" name="user"/></td>
                </tr>
```

```
                <tr>
                    <td><input type="submit" value="提交"/></td>
                </tr>
            </table>
        </form>
    </body>
</html>
```

代码模板 UseSessionServlet.java 如下：

```
package servlet;
import java.io.*;
import javax.servlet.*;
import javax.servlet.http.*;
public class UseSessionServlet extends HttpServlet {
    public void init(ServletConfig config) throws ServletException{
        super.init(config);
    }
    public void doPost(HttpServletRequest request, HttpServletResponse
            response) throws ServletException, IOException{
        response.setContentType("text/html;charset=GBK");
        PrintWriter out=response.getWriter();
        request.setCharacterEncoding("GBK");
        String name=request.getParameter("user");
        HttpSession session=request.getSession(true);   //获得用户的会话对象
        session.setAttribute("myName", name);
        out.println("<htm><body>");
        out.println("您请求的 servlet 对象是: "+getServletName());
        out.println("<br>您的会话 ID 是: "+session.getId());
        out.println("<br>请单击请求另一个 servlet: ");
        out.println("<br><a href=showMyName>请求另一个 servlet</a>");
        out.println("</body></htm>");
    }
    public void doGet(HttpServletRequest request, HttpServletResponse
            response) throws ServletException, IOException{
        doPost(request, response);
    }
}
```

代码模板 ShowNameServlet.java 如下：

```
package servlet;
import java.io.*;
import javax.servlet.*;
import javax.servlet.http.*;
public class ShowNameServlet extends HttpServlet {
```

```
    public void init(ServletConfig config) throws ServletException{
        super.init(config);
    }
    public void doPost(HttpServletRequest request, HttpServletResponse
            response) throws ServletException, IOException{
        response.setContentType("text/html;charset=GBK");
        PrintWriter out=response.getWriter();
        HttpSession session=request.getSession(true);  //获得用户的会话对象
        String name=(String)session.getAttribute("myName");
        out.println("<htm><body>");
        out.println("您请求的 servlet 对象是: "+getServletName());
        out.println("<br>您的会话 ID 是: "+session.getId());
        out.println("<br>您的会话中存储的用户名是: "+name);
        out.println("</body></htm>");
    }
    public void doGet(HttpServletRequest request, HttpServletResponse
            response) throws ServletException, IOException{
        doPost(request, response);
    }
}
```

用户的会话对象 session 可以在 JSP 页面中不用声明直接使用,而在 Servlet 类中必须先使用 request 对象获得用户的会话对象,然后再使用它。

7.6.3 拓展训练

1. servlet 对象如何获得用户的会话对象?

2. 请阐述在 JSP 页面中使用会话对象 session 和在 servlet 中使用会话对象 session 有什么不同,并举例说明。

3. 假设创建 servlet 的类是 my. servlet. MyFirstServlet,创建的 servlet 对象的名字是 first,如果使用表单请求该 servlet,表单的 action 的值是 isgo。应该怎么为该 servlet 编写部署文件 web. xml?

7.7 小 结

- Java Servlet 技术的核心思想就是在服务器端创建能响应用户请求的对象,该对象习惯上被称为一个 servlet 对象。
- 要想让 Web 服务器用 Servlet 类编译后的字节码文件创建 servlet 对象,必须为 Web 服务器编写一个部署文件 web. xml。
- servlet 对象第一次被请求加载时,服务器创建一个 servlet 对象,该对象调用 init 方法完成必要的初始化工作。当 servlet 对象成功创建后,servlet 对象就调用 service

方法来处理用户的请求并返回响应。

- 当服务器接收到一个 servlet 请求时,就会产生一个新线程,在这个线程中让 servlet 对象调用 service 方法为请求作出响应。实际上,service 方法首先检查 HTTP 请求类型(get、post 等),并在 service 方法中根据用户的请求方式,对应地再调用 doGet 或 doPost 方法。
- 重定向的功能是将用户从当前 JSP 页面或 servlet 定向到另一个 JSP 页面或 servlet,以前的 request 中存放的变量全部失效,并进入一个新的 request 作用域;转发的功能是将用户对当前 JSP 页面或 servlet 的请求转发给另一个 JSP 页面或 servlet,以前的 request 中存放的变量不会失效。
- 一个用户在同一个 Web 服务中的不同 servlet 中获取的 session 对象是完全相同的,但是不同用户的 session 对象互不相同。

第8章

基于Servlet的MVC模式

前面的章节中已经学习了JSP和Servlet技术，使用它们可以开发出完整的Web应用程序。但有时会把大量的Java代码写在JSP页面中，把HTML代码写在Servlet中。这样会造成代码编写不容易，日后维护也不容易。因此，学习Web应用程序的设计模式就显得尤为重要了。本章就将学习一种非常典型的Web应用程序的设计模式——基于Servlet的MVC模式。

8.1 JSP中的MVC模式

8.1.1 知识要点

1. MVC的概念

MVC是Model、View、Controller的缩写，分别代表Web应用程序中的三种职责：

- 模型——用于存储数据以及处理用户请求的业务逻辑。
- 视图——向控制器提交数据，显示模型中的数据。
- 控制器——根据视图提出的请求，判断将请求和数据交给哪个模型处理，处理后的有关结果交给哪个视图更新显示。

2. JSP中的MVC模式

JSP中的MVC模式的具体实现如下：

- 模型。一个或多个JavaBean对象，用于存储数据(实体模型，由JavaBean类创建)和处理业务逻辑(业务模型，由一般的Java类创建)。
- 视图。一个或多个JSP页面，向控制器提交数据和为模型提供数据显示，JSP页面主要使用HTML标记和JavaBean标记来显示数据。
- 控制器。一个或多个servlet对象，根据视图提交的请求进行控制，即把请求转发给处理业务逻辑的JavaBean，并将处理结果存放到实体模型JavaBean中，输出给视图显示。

JSP中的MVC模式的流程如图8-1所示。

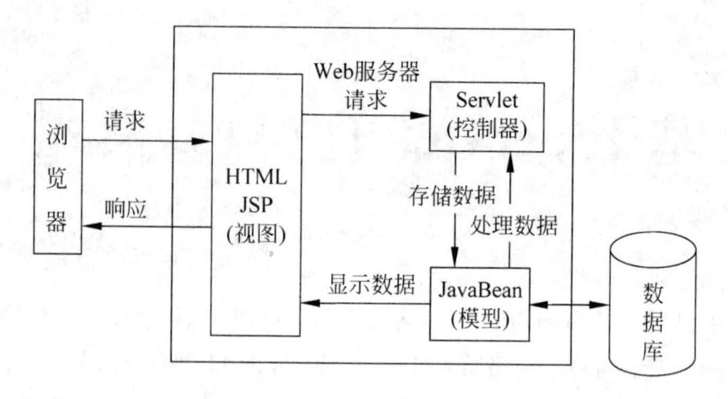

图 8-1　JSP 中的 MVC 模式

8.1.2　技能操作

运用 MVC 模式，实现登录验证过程。具体任务如下：

编写一个简单的 Web 应用程序：用户登录验证程序。视图 login.jsp 提交数据请求(用户名和密码)；控制器 controllerServlet(ControllerServlet 类负责创建)接收请求信息，然后把请求信息封装在 user(UserBean 类负责创建)实体模型中，控制器把 user 模型传递给 userCheck 业务模型去处理(UserCheckBean 类负责创建)；如果用户名和密码输入正确，则返回 success.jsp 页面，否则返回 login.jsp 页面。程序运行效果如图 8-2 所示。

http://localhost:8080/ch08/login.jsp

欢迎登录本系统，请输入用户名与密码!

用户名：

密　码：

提交　重置

http://localhost:8080/ch08/isLogin

恭喜JSPMVC登录成功!

图 8-2　登录验证程序

代码模板 web.xml 如下：

```
<?xml version="1.0" encoding="UTF-8"?>
<web-app>
<servlet>
    <servlet-name>controllerServlet</servlet-name>
    <servlet-class>servlet.ControllerServlet</servlet-class>
</servlet>
<servlet-mapping>
    <servlet-name>controllerServlet</servlet-name>
    <url-pattern>/isLogin</url-pattern>
```

```
</servlet-mapping>
</web-app>
```

代码模板 UserBean.java(实体模型)如下：

```
package entity.bean;
public class UserBean {
    String name;
    String pwd;
    public UserBean(){
    }
    public String getName() {
        return name;
    }
    public void setName(String name) {
        this.name = name;
    }
    public String getPwd() {
        return pwd;
    }
    public void setPwd(String pwd) {
        this.pwd = pwd;
    }
}
```

代码模板 UserCheckBean.java(业务模型)如下：

```
package busynees.bean;
import entity.bean.UserBean;
public class UserCheckBean {
    //验证登录
    public boolean validate(UserBean user){
        if(user!=null&&user.getName().equals("JSPMVC")){
            if(user.getPwd().equals("MVC")){
                return true;
            }
            return false;
        }
        return false;
    }
}
```

代码模板 login.jsp(视图)如下：

```
<%@ page language="java" contentType="text/html; charset=GBK" pageEncoding=
"GBK"%>
<html>
```

```
        <head>
            <title>login.jsp</title>
        </head>
        <body>
            <form action="isLogin" method="post">
            欢迎登录本系统,请输入用户名与密码!
            <table>
                <tr>
                    <td>用户名: </td>
                    <td><input type="text" name="name"/></td>
                </tr>
                <tr>
                    <td>密码: </td>
                    <td><input type="password" name="pwd"/></td>
                </tr>
                <tr>
                    <td><input type="submit" value="提交"/></td>
                    <td><input type="reset" value="重置"/></td>
                </tr>
            </table>
            </form>
        </body>
</html>
```

代码模板 success.jsp(视图)如下:

```
<%@ page language = "java" contentType = "text/html; charset = GBK" pageEncoding =
"GBK"%>
<%@ page import="entity.bean.UserBean" %>
<html>
<head>
<title>success.jsp</title>
</head>
<body>
<jsp:useBean id="userBean" type="entity.bean.UserBean" scope="request"/>
恭喜<jsp:getProperty property="name" name="userBean"/>登录成功!
</body>
</html>
```

代码模板 ControllerServlet.java(控制器)如下:

```
package servlet;
import java.io.IOException;
import java.io.PrintWriter;
import javax.servlet.RequestDispatcher;
import javax.servlet.ServletConfig;
```

```
import javax.servlet.ServletException;
import javax.servlet.http.HttpServlet;
import javax.servlet.http.HttpServletRequest;
import javax.servlet.http.HttpServletResponse;
import busynees.bean.UserCheckBean;
import entity.bean.UserBean;
public class ControllerServlet extends HttpServlet {
    public void init(ServletConfig config) throws ServletException{
        super.init(config);
    }
    public void doPost(HttpServletRequest request, HttpServletResponse response)
            throws IOException, ServletException{
        request.setCharacterEncoding("GBK");                //设置编码,防止中文乱码
        String name=request.getParameter("name");           //获取客户提交的信息
        String password=request.getParameter("pwd");        //获取客户提交的信息
        UserBean user=new UserBean();                       //实例化实体模型 user
        user.setName(name);                                 //把数据存在模型 user 中
        user.setPwd(password);                              //把数据存在模型 user 中
        UserCheckBean userCheck=new UserCheckBean();        //实例化业务模型 userCheck
        if(userCheck.validate(user)){
            request.setAttribute("userBean", user);         //把装有数据的实体模型 user
            RequestDispatcher dis=request.getRequestDispatcher("success.jsp");
            dis.forward(request, response);
        }else{
            response.sendRedirect("login.jsp");
        }
    }
    public void doGet(HttpServletRequest request, HttpServletResponse response)
            throws IOException, ServletException{
        doPost(request, response);
    }
}
```

在 JSP 的 MVC 模式中，控制器 servlet 创建的实体模型 JavaBean 也涉及到生命周期，生命周期分别为 request、session 和 application。下面以技能操作任务中的实体模型 user 来讨论这三种生命周期的模型的用法。

- request 周期的模型。

使用 request 周期的模型一般需要以下几个环节：

① 创建模型并把数据保存到模型中。

在 servlet 中需要这样的代码：

```
UserBean user=new UserBean();                               //实例化模型 user
user.setName(name);                                         //把数据存在模型 user 中
user.setPwd(password);                                      //把数据存在模型 user 中
```

② 将模型保存到 request 对象中并转发给视图 JSP。

在 servlet 中需要这样的代码：

```
request.setAttribute("userBean", user);      //把装有数据的模型 user 输出给视图 success.jsp 页面
RequestDispatcher dis=request.getRequestDispatcher("success.jsp");
dis.forward(request, response);
```

request.setAttribute("userBean", user)这句代码指定了查找 JavaBean 的关键字，并决定了 JavaBean 的生命周期为 request。

③ 视图更新。

servlet 所转发的页面，例如 success.jsp 页面，必须使用 useBean 标记获得 servlet 所创建的 JavaBean 对象（视图不负责创建 JavaBean）。在 JSP 页面需要使用这样的代码：

```
<jsp:useBean id="userBean" type="entity.bean.UserBean" scope="request"/>
<jsp:getProperty property="name" name="userBean"/>
```

标记中的 id 就是 servlet 所创建的模型 JavaBean，它和 request 对象中的关键字对应。因为在视图中不创建 JavaBean 对象，所以在 useBean 标记中使用 type 属性，而不使用 class 属性。useBean 标记中的 scope 必须和存储模型时的范围（request）一致。

- session 周期的模型。

使用 session 周期的模型一般需要以下几个环节：

① 创建模型并把数据保存到模型中。

在 servlet 中需要这样的代码：

```
UserBean user=new UserBean();                //实例化模型 user
user.setName(name);                          //把数据存在模型 user 中
user.setPwd(password);                       //把数据存在模型 user 中
```

② 将模型保存到 session 对象中并转发给视图 JSP。

在 servlet 中需要这样的代码：

```
session.setAttribute("userBean", user);      //把装有数据的模型 user 输出给视图
                                               success.jsp 页面
RequestDispatcher dis=request.getRequestDispatcher("success.jsp");
dis.forward(request, response);
```

session.setAttribute("userBean", user)这句代码指定了查找 JavaBean 的关键字，并决定了 JavaBean 的生命周期为 session。

③ 视图更新。

servlet 所转发的页面，例如 success.jsp 页面，必须使用 useBean 标记获得 servlet 所创建的 JavaBean 对象（视图不负责创建 JavaBean）。在 JSP 页面需要使用这样的代码：

```
<jsp:useBean id="userBean" type="entity.bean.UserBean" scope="session"/>
<jsp:getProperty property="name" name="userBean"/>
```

标记中的 id 就是 servlet 所创建的模型 JavaBean，它和 session 对象中的关键字对应。因为在视图中不创建 JavaBean 对象，所以在 useBean 标记中使用 type 属性，而不使用 class 属性。useBean 标记中的 scope 必须和存储模型时的范围(session)一致。

注意：对于生命周期为 session 的模型，servlet 不仅可以使用 RequestDispatcher 对象转发给 JSP 页面，也可以使用 response 的重定向方法(sendRedirect)定向到 JSP 页面。

- application 周期的模型。

使用 application 周期的模型一般需要以下几个环节：

① 创建模型并把数据保存到模型中。

在 servlet 中需要这样的代码：

```
UserBean user=new UserBean();                    //实例化模型 user
user.setName(name);                              //把数据存在模型 user 中
user.setPwd(password);                           //把数据存在模型 user 中
```

② 将模型保存到 application 对象中并转发给视图 JSP。

在 servlet 中需要这样的代码：

```
application.setAttribute("userBean", user); //把装有数据的模型 user 输出给视图 success.jsp 页面
RequestDispatcher dis=request.getRequestDispatcher("success.jsp");
dis.forward(request, response);
```

application.setAttribute("userBean", user)这句代码指定了查找 JavaBean 的关键字，并决定了 JavaBean 的生命周期为 application。

③ 视图更新。

servlet 所转发的页面，例如 success.jsp 页面，必须使用 useBean 标记获得 servlet 所创建的 JavaBean 对象(视图不负责创建 JavaBean)。在 JSP 页面需要使用这样的代码：

```
<jsp:useBean id="userBean" type="entity.bean.UserBean" scope="application"/>
<jsp:getProperty property="name" name="userBean"/>
```

标记中的 id 就是 servlet 所创建的模型 JavaBean，它和 application 对象中的关键字对应。因为在视图中不创建 JavaBean 对象，所以在 useBean 标记中使用 type 属性，而不使用 class 属性。useBean 标记中的 scope 必须和存储模型时的范围(application)一致。

注意：对于生命周期为 application 的模型，servlet 不仅可以使用 RequestDispatcher 对象转发给 JSP 页面，也可以使用 response 的重定向方法(sendRedirect)定向到 JSP 页面。

8.1.3　拓展训练

1. 以下不属于 MVC 设计模式中 3 个模块的是(　　)。

   A. 模型　　　B. 表示层　　　C. 视图　　　D. 控制器

2. 在 MVC 模式中，(　　)用于客户端应用程序的图形数据表示，与实际数据处理无关。

A. 模型　　B. 视图　　C. 控制器　　D. 数据

3. 在 MVC 设计模式中，(　　)接收用户请求数据。

A. HTML　　B. JSP　　C. Servlet　　D. 业务类

4. 将 8.1.2 节中的实体模型 user 的生命周期改为 session，并运行程序。

8.2　使用 MVC 模式查询数据库

8.2.1　知识要点

使用 MVC 模式设计 Web 应用时，尽量把实体模型与业务模型分开实现，以方便以后维护。例如，在使用 MVC 模式查询数据库这个 Web 应用中数据的封装由实体模型 Goods 完成，而处理数据由业务模型 GoodsDao 完成和数据库的操作由业务模型 DataBaseBean 完成。

8.2.2　技能操作

使用 MVC 模式设计 Web 应用。具体任务如下：

设计一个 Web 应用，有两个 JSP 页面(addGoods.jsp 和 showAllGoods.jsp)、三个 JavaBean(实体模型 Goods、业务模型 GoodsDao 和业务模型 DataBaseBean)以及一个 servlet(GoodsControllerServlet)。用户在 JSP 页面 addGoods.jsp 中输入商品的信息，提交给 servlet，该 servlet 负责添加商品(调用业务模型 GoodsDao 的 addGoods 方法)、查询商品(调用业务模型 GoodsDao 的 getAllGoods 方法)，并把查询结果显示在 showAllGoods.jsp 页面中。Web 应用中使用了 6.2 节曾经使用过的数据表 goodsInfo。

页面运行效果如图 8-3 和图 8-4 所示。

http://localhost:8080/ch08/addGoods.jsp

商品编号是主键，由程序自动产生！

商品名称：蛋卷
商品价格：18.5
商品类型：食品
添加　重置

图 8-3　商品信息输入页面

http://localhost:8080/ch08/addServlet

商品编号	商品名称	商品价格	商品类别
1	牙膏	12.0	日用品
2	冰箱	2500.0	电器
12	蛋卷	18.5	食品
4	苹果	48.0	水果
5	上衣	1800.0	服装
6	书包	78.0	文具
7	香蕉	17.0	水果
10	豆浆机	299.99	电器
11	蛋卷	18.5	食品

图 8-4　商品信息显示页面

要想达到上述效果，需要为web.xml文件添加如下子标记：

```
<servlet>
    <servlet-name>addServlet</servlet-name>
    <servlet-class>servlet.GoodsControllerServlet</servlet-class>
</servlet>
<servlet-mapping>
    <servlet-name>addServlet</servlet-name>
    <url-pattern>/addServlet</url-pattern>
</servlet-mapping>
```

代码模板Goods.java(实体模型)如下：

```
package entity.bean;
public class Goods {
    int goodsId;
    String goodsName;
    double goodsPrice;
    String goodsType;
    public Goods(){
    }
    public int getGoodsId() {
        return goodsId;
    }
    public void setGoodsId(int goodsId) {
        this.goodsId = goodsId;
    }
    public String getGoodsName() {
        return goodsName;
    }
    public void setGoodsName(String goodsName) {
        this.goodsName = goodsName;
    }
    public double getGoodsPrice() {
        return goodsPrice;
    }
    public void setGoodsPrice(double goodsPrice) {
        this.goodsPrice = goodsPrice;
    }
    public String getGoodsType() {
        return goodsType;
    }
    public void setGoodsType(String goodsType) {
        this.goodsType = goodsType;
    }
}
```

代码模板 GoodsDao.java(业务模型)如下：

```
package busynees.bean;
import java.sql.*;
import java.util.ArrayList;
import entity.bean.Goods;
public class GoodsDao {
    //获得新添加的商品编号
    public synchronized int getID(){
        Connection con=DataBaseBean.getCon();
        PreparedStatement ps=null;
        ResultSet rs=null;
        int id=0;
        try {
            ps=con.prepareStatement("select max(goodsId) from goodsInfo");
            rs=ps.executeQuery();
            if(rs.next()){
                id=rs.getInt(1)+1;
            }
        } catch (SQLException e) {
            // TODO Auto-generated catch block
            e.printStackTrace();
        }finally{
            DataBaseBean.close(rs);
            DataBaseBean.close(ps);
            DataBaseBean.close(con);
        }
        return id;
    }
    //添加商品
    public boolean addGoods(Goods goods){
        Connection con=DataBaseBean.getCon();
        PreparedStatement ps=null;
        try {
            ps=con.prepareStatement("insert into goodsInfo values(?,?,?,?)");
            ps.setInt(1, goods.getGoodsId());
            ps.setString(2, goods.getGoodsName());
            ps.setDouble(3, goods.getGoodsPrice());
            ps.setString(4, goods.getGoodsType());
            int i=ps.executeUpdate();
            if(i>0)
                return true;
        } catch (SQLException e) {
            // TODO Auto-generated catch block
            e.printStackTrace();
        }finally{
```

```
            DataBaseBean.close(ps);
            DataBaseBean.close(con);
        }
        return false;
    }
    //查询商品
    public ArrayList<Goods> getAllGoods(){
        Connection con=DataBaseBean.getCon();
        PreparedStatement ps=null;
        ResultSet rs=null;
        ArrayList<Goods> al=new ArrayList<Goods>();
        try {
            ps=con.prepareStatement("select * from goodsInfo");
            rs=ps.executeQuery();
            while(rs.next()){
                Goods gd=new Goods();
                gd.setGoodsId(rs.getInt(1));
                gd.setGoodsName(rs.getString(2));
                gd.setGoodsPrice(rs.getDouble(3));
                gd.setGoodsType(rs.getString(4));
                al.add(gd);
            }
        } catch (SQLException e) {
            // TODO Auto-generated catch block
            e.printStackTrace();
        }finally{
            DataBaseBean.close(rs);
            DataBaseBean.close(ps);
            DataBaseBean.close(con);
        }
        return al;
    }
}
```

代码模板 DataBaseBean.java(数据库业务模型)如下：

```
package busynees.bean;
import java.sql.Connection;
import java.sql.DriverManager;
import java.sql.PreparedStatement;
import java.sql.ResultSet;
import java.sql.SQLException;
public class DataBaseBean {
    public static Connection getCon(){
        Connection con=null;
        try {
```

```
            Class.forName("oracle.jdbc.driver.OracleDriver");
        } catch (ClassNotFoundException e) {
            e.printStackTrace();
        }
        try {
          con = DriverManager.getConnection("jdbc:oracle:thin:@127.0.0.1:1521:orcl",
"system","system");
        } catch (SQLException e) {
            e.printStackTrace();
        }
        return con;
    }
    public static void close(ResultSet rs){
        if(rs!=null){
            try {
                rs.close();
            } catch (SQLException e) {
                e.printStackTrace();
            }
        }
    }
    public static void close(PreparedStatement ps){
        if(ps!=null){
            try {
                ps.close();
            } catch (SQLException e) {
                e.printStackTrace();
            }
        }
    }
    public static void close(Connection con){
        if(con!=null){
            try {
                con.close();
            } catch (SQLException e) {
                e.printStackTrace();
            }
        }
    }
}
```

代码模板 addGoods.jsp(视图 1)如下：

```
<%@ page language="java" contentType="text/html; charset=GBK" pageEncoding=
"GBK"%>
<html>
```

```
<head>
<title>addGoods.jsp</title>
</head>
<body>
    <h4>商品编号是主键,由程序自动产生!</h4>
    <form action="addServlet" method="post">
    <table border="1">
        <tr>
            <td>商品名称:</td>
            <td><input type="text" name="goodsName"/></td>
        </tr>

        <tr>
            <td>商品价格:</td>
            <td><input type="text" name="goodsPrice"/></td>
        </tr>

        <tr>
            <td>商品类型:</td>
            <td>
                <select name="goodsType">
                    <option value="日用品">日用品
                    <option value="电器">电器
                    <option value="食品">食品
                    <option value="水果">水果
                    <option value="服装">服装
                    <option value="文具">文具
                    <option value="其他">其他
                </select>
            </td>
        </tr>

        <tr>
            <td><input type="submit" value="添加"></td>
            <td><input type="reset" value="重置"></td>
        </tr>
    </table>
    </form>
</body>
</html>
```

代码模板 showAllGoods.jsp(视图 2)如下：

```
<%@ page language="java" contentType="text/html; charset=GBK" pageEncoding=
"GBK"%>
<%@ page import="java.util.ArrayList" %>
```

```
<%@ page import="entity.bean.Goods" %>
<html>
<head>
<title>showAllGoods.jsp</title>
</head>
<body>
<%
    ArrayList<Goods> al=(ArrayList<Goods>)request.getAttribute("goods");
    if(al!=null&&al.size()>0){
%>
    <table border=1>
        <tr>
            <th>商品编号</th>
            <th>商品名称</th>
            <th>商品价格</th>
            <th>商品类别</th>
        </tr>
    <%
        for(int i=0;i<al.size();i++){
            Goods good=al.get(i);
    %>
        <tr>
            <td><%=good.getGoodsId() %></td>
            <td><%=good.getGoodsName() %></td>
            <td><%=good.getGoodsPrice() %></td>
            <td><%=good.getGoodsType() %></td>
        </tr>
    <%
        }
    %>
    </table>
<%
    }else{
        out.print("还没有商品");
    }
%>
</body>
</html>
```

代码模板 GoodsControllerServlet.java(控制器)如下:

```
package servlet;
import java.io.IOException;
import java.util.ArrayList;
import javax.servlet.RequestDispatcher;
import javax.servlet.ServletConfig;
```

```
import javax.servlet.ServletException;
import javax.servlet.http.HttpServlet;
import javax.servlet.http.HttpServletRequest;
import javax.servlet.http.HttpServletResponse;
import busynees.bean.GoodsDao;
import entity.bean.Goods;
public class GoodsControllerServlet extends HttpServlet {
    public void init(ServletConfig config) throws ServletException{
        super.init(config);
    }
    public void doPost(HttpServletRequest request,HttpServletResponse response)
            throws IOException,ServletException{
        request.setCharacterEncoding("GBK");              //设置编码,防止中文乱码
        String goodsName=request.getParameter("goodsName"); //获得视图提交的信息
        String goodsPrice=request.getParameter("goodsPrice");
        String goodsType=request.getParameter("goodsType");
        Goods goods=new Goods();                          //创建实体模型 goods
        goods.setGoodsId(gd.getID());                     //把数据存储到实体模型中
        goods.setGoodsName(goodsName);                    //把数据存储到实体模型中
        goods.setGoodsPrice(Double.parseDouble(goodsPrice)); //把数据存储到实体模型中
        goods.setGoodsType(goodsType);                    //把数据存储到实体模型中
        GoodsDao gd= new GoodsDao();                      //创建业务模型 gd
        if(gd.addGoods(goods)){ //业务模型 gd 调用 addGoods 方法添加商品
            ArrayList<Goods> al= gd.getAllGoods(); //业务模型 gd 调用 getAllGoods 方法查询商品
            request.setAttribute("goods", al);            //把商品数组保存到 request 里面
            RequestDispatcher dis=request.getRequestDispatcher("showAllGoods.jsp");
            dis.forward(request, response);
        }else{
            response.sendRedirect("addGoods.jsp");
        }
    }
    public void doGet(HttpServletRequest request,HttpServletResponse response)
            throws IOException,ServletException{
        doPost(request,response);
    }
}
```

使用MVC模式设计Web应用时控制器尽量不处理数据，它只是起到控制转发的功能，而数据交给业务模型处理，如技能操作任务中的控制器。

8.2.3 拓展训练

1. 使用MVC模式设计Web应用有什么好处？
2. MVC中的模型是由servlet负责创建，还是由JSP页面负责创建？
3. 使用MVC模式设计一个Web应用，用户通过JSP页面inputNumber.jsp输入两个

操作数,并选择一种运算符,单击“提交”按钮后,调用 HandleComputer. java 这个 servlet。在 HandleComputer. java 中获取用户输入的数字和运算符并将这些内容放入 ComputerBean. java 这个 JavaBean 中,在 showResult. jsp 中调用 JavaBean 显示计算的结果。

8.3 小　结

MVC 模式将“视图”、“模型”和“控制器”有效地组合。在 JSP 的 MVC 模式中,视图是一个或多个 JSP 页面,作用是向控制器提交数据和为模型提供数据显示;模型是一个或多个 JavaBean 对象,用于存储数据(实体模型,由 JavaBean 类创建)和处理业务逻辑(业务模型,由一般的 Java 类创建);控制器是一个或多个 servlet 对象,根据视图提交的请求进行控制,即把请求转发给处理业务逻辑的 JavaBean,并将处理结果存放到实体模型 JavaBean 中,输出给视图显示。

第9章

开发 Web 应用过滤器

在 Web 开发过程中，可能有这样的需求：某些页面只希望几个特定的用户浏览。对于这样的访问权限的控制，该如何实现呢？过滤器(Filter)就可以完成上述需求的实现。过滤器作用于服务器(Servlet)处理请求之前或服务器(Servlet)响应请求之前。也就是说，它既可以过滤浏览器对服务器的请求，也可以过滤服务器对浏览器的响应，如图 9-1 所示。

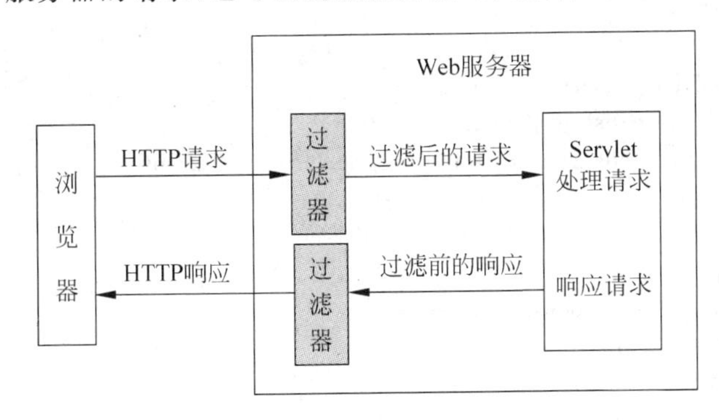

图 9-1　过滤器

如何编写过滤器类，又如何使用过滤器类，关于这些将在本章重点介绍。

9.1　Filter 类与 filter 对象

9.1.1　知识要点

编写一个过滤器类很简单，只要实现 javax. servlet 包中的 Filter 接口。实现 Filter 接口的类习惯上称为一个 Filter 类，这样的类创建的对象又称为 filter 对象。

9.1.2　技能操作

编写一个类实现 Filter 接口，该类的主要功能是创建过滤器。具体任务如下：新建一个 Web 工程 ch09，在该 Web 工程中编写一个简单的 Filter 类 MyFirstFilter，Filter 类实现如下功能：

不管用户请求该 Web 工程的哪个页面或 servlet，都会在浏览器中先出现“首先执行过滤器"这样的响应信息。

代码模板 MyFirstFilter.java 如下：

```
package filters;
import java.io.IOException;
import java.io.PrintWriter;
import javax.servlet.Filter;
import javax.servlet.FilterChain;
import javax.servlet.FilterConfig;
import javax.servlet.ServletException;
import javax.servlet.ServletRequest;
import javax.servlet.ServletResponse;
public class MyFirstFilter implements Filter {                //实现 Filter 接口的类
    public void destroy() {
    }
    public void doFilter(ServletRequest request,
            ServletResponse response,
            FilterChain chain) throws IOException, ServletException {
        //设置响应类型
        response.setContentType("text/html;charset=GBK");
        //获得输出对象 out
        PrintWriter out=response.getWriter();
        //在浏览器中输出
        out.print("首先执行过滤器<br>");
        //执行下一个过滤器
        chain.doFilter(request, response);
    }
    public void init(FilterConfig fConfig) throws ServletException {
    }
}
```

可以从任务中 MyFirstFilter 类的源代码看出：Filter 接口与 Servlet 接口很类似，同样都有 init()与 destroy()方法，还有一个 doFilter()方法类似于 Servlet 接口的 service()方法。下面分别介绍这三个方法的功能：

(1) public void init(FilterConfig Config) throws ServletException

方法的功能是初始化过滤器对象。方法中的参数 fConfig 是 FilterConfig 的对象，该对象代表 web.xml 中为过滤器定义的对象。如果在 web.xml 中为过滤器设置了初始参数，则可以通过 FilterConfig 的 getInitParameter(String paramName)方法获得初始参数值。

(2) public void doFilter(ServletRequest request, ServletResponse response, FilterChain chain) throws IOException, ServletException

当 Web 服务器使用 servlet 对象调用 service()方法处理请求前，发现应用了某个过滤器时，Web 服务器就会自动调用该过滤器的 doFilter()方法。在 doFilter()方法中有这样一条语句：

```
chain.doFilter(request, response);
```

如果执行了该语句，就会执行下一个过滤器(根据<filter-mapping>在web.xml中出现的先后顺序执行过滤器)；如果没有下一个过滤器，就返回请求目标程序。如果因为某个原因没有执行"chain.doFilter(request, response);"，则请求就不会继续交给以后的过滤器或请求目标程序，这就是所谓的拦截请求。

(3) public void destroy()

当Web服务器终止服务时，destroy方法会被执行，使filter对象消亡。

编写完Filter类的源文件，并编译了源文件，这时Web服务器是不是就可以运行filter对象呢？不可以，需要部署filter之后，Web服务器才可以运行filter对象。

9.1.3 拓展训练

尝试找一下技能操作任务中的Filter类编译后的字节码文件。

9.2 filter对象的部署与运行

9.2.1 知识要点

要想让Web服务器用Filter类编译后的字节码文件创建filter对象，必须在Web工程的web.xml文件里部署自己的filter。

9.2.2 技能操作

部署和运行过滤器。具体任务如下：

- 部署filter。
- 运行filter。

1. 部署filter

为了在web.xml文件里部署9.1.2节中的MyFirstFilter，需要在web.xml文件里找到<web-app></web-app>标记，然后在<web-app></web-app>标记中添加如下内容：

```
<filter>
    <filter-name>myFirstFilter</filter-name>
    <filter-class>filters.MyFirstFilter</filter-class>
</filter>
<filter-mapping>
    <filter-name>myFirstFilter</filter-name>
    <url-pattern>/*</url-pattern>
```

```
</filter-mapping>
```

为了更好地理解上述代码，现将<filter>标记及其子标记、<filter-mapping>标记及其子标记相关内容说明如下。

1）<filter>标记及其子标记

web.xml 文件中可以有若干个<filter>标记，该标记的内容由 Web 服务器负责处理。<filter>标记中有两个子标记：<filter-name>和<filter-class>，其中<filter-name>子标记的内容是 Web 服务器创建的 filter 对象的名字。web.xml 文件中可以有若干个<filter>标记，但要求它们的<filter-name>子标记的内容互不相同。<filter-class>子标记的内容指定 Web 服务器用哪个类来创建 filter 对象，如果 filter 对象已经创建，那么 Web 服务器就不再使用指定的类创建。

如果在过滤器初始化时，需要读取一些参数的值，则可以在<filter>标记中使用<init-param>子标记设置。例如：

```
<filter>
    <filter-name>myFirstFilter</filter-name>
    <filter-class>filters.MyFirstFilter</filter-class>
    <init-param>
        <param-name>encoding</param-name>
        <param-value>GBK</param-value>
    </init-param>
</filter>
```

那么就可以在 filter 的 init() 方法中使用参数 fConfig(FilterConfig 的对象)调用 FilterConfig 的 getInitParameter(String paramName)方法获得参数值。例如：

```
public void init(FilterConfig fConfig) throws ServletException{
    String en= fConfig.getInitParameter("encoding");
}
```

2）<filter-mapping>标记及其子标记

web.xml 文件中出现一个<filter>标记就会对应地出现一个<filter-mapping>标记。<filter-mapping>标记中也有两个子标记：<filter-name>和<url-pattern>。其中<filter-name>子标记的内容是 Web 服务器创建的 filter 对象的名字(该名字必须和<filter>标记的子标记<filter-name>的内容相同)；<url-pattern>子标记用来指定用户用怎样的模式请求 filter 对象。

如果有某个 URL 或 servlet 会应用多个过滤器，则根据<filter-mapping>标记在 web.xml 中出现的先后顺序执行过滤器。

2. 运行 filter

只要用户请求的 URL 和<filter-mapping>的子标记<url-pattern>指定的模式匹配，

Web 服务器就会自动调用该 filter 的 doFilter()方法。如 9.1.2 节中的 MyFirstFilter 过滤器在 web.xml 中的<url-pattern>指定值为/＊,"/＊"代表任何页面或 servlet 的请求。

9.2.3　拓展训练

按照本节的任务内容将 9.1.2 节中的过滤器 MyFirstFilter 部署成功,并运行 Web 应用程序测试该过滤器。

9.3　创建 Web 应用过滤器

9.3.1　知识要点

在 Web 工程中,某些页面或 servlet 只有用户登录成功才能访问。直接在应用程序每个相关的源代码中判断用户是否登录成功,这并不是科学的做法。我们可以实现一个登录验证过滤器,在 Web 工程的 web.xml 中设置并使用该过滤器,就可以不用在每个相关的源代码中验证用户是否登录成功。

9.3.2　技能操作

创建 Web 应用之登录验证过滤器。具体任务如下：

新建一个 Web 工程 loginValidate,在该 Web 工程中至少编写两个 JSP 页面(login.jsp 与 loginSuccess.jsp)和一个 servlet(由 LoginServlet.java 负责创建)。用户在 login.jsp 页面中输入用户名和密码后,提交给 servlet,在 servlet 中判断用户名和密码是否正确,若正确,则跳转到 loginSuccess.jsp；若错误,则回到 login.jsp 页面。但该 Web 工程有另外一个要求：除了访问 login.jsp 页面外,别的页面或 servlet 都不能直接访问,必须先登录成功才能访问。我们在设计这个 Web 工程时,编写了一个登录验证过滤器并在该 Web 工程中使用。

页面运行效果如图 9-2 至图 9-4 所示。

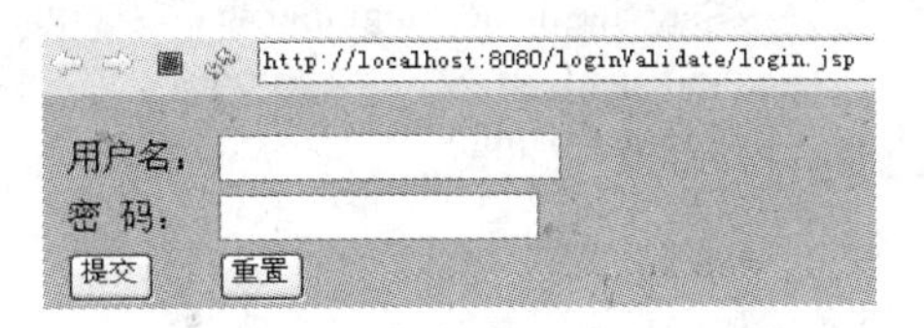

图 9-2　登录画面

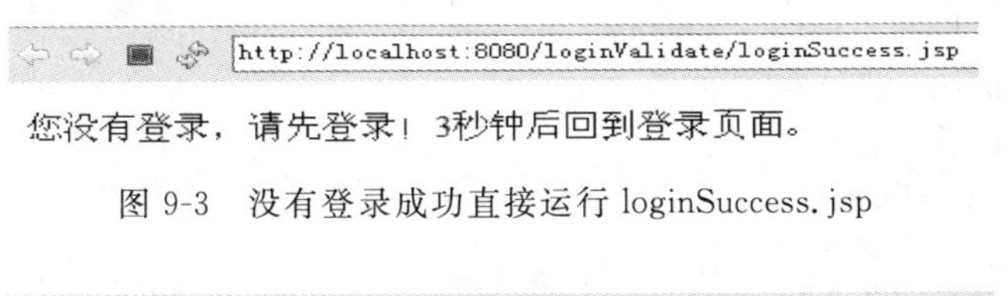

图 9-3　没有登录成功直接运行 loginSuccess.jsp

http://localhost:8080/loginValidate/loginSuccess.jsp

恭喜filter登录成功！

图 9-4　登录成功页面

代码模板 web. xml 如下：

```
<web-app>

  <filter>
    <filter-name>loginValidateFilter</filter-name>
    <filter-class>filters.LoginFilter</filter-class>
    <init-param>
      <param-name>login_uri</param-name>
      <param-value>/login.jsp</param-value>
    </init-param>
    <init-param>
      <param-name>login_Servlet</param-name>
      <param-value>/isLogin</param-value>
    </init-param>
  </filter>

  <filter-mapping>
    <filter-name>loginValidateFilter</filter-name>
    <url-pattern>/*</url-pattern>
  </filter-mapping>

  <servlet>
    <servlet-name>loginServlet</servlet-name>
    <servlet-class>servlet.LoginServlet</servlet-class>
  </servlet>

  <servlet-mapping>
    <servlet-name>loginServlet</servlet-name>
    <url-pattern>/isLogin</url-pattern>
  </servlet-mapping>

</web-app>
```

代码模板 LoginFilter. java（过滤器）如下：

```
package filters;
import java.io.IOException;
import java.io.PrintWriter;
import javax.servlet.Filter;
import javax.servlet.FilterChain;
import javax.servlet.FilterConfig;
import javax.servlet.ServletException;
import javax.servlet.ServletRequest;
import javax.servlet.ServletResponse;
import javax.servlet.http.HttpServletRequest;
```

```
import javax.servlet.http.HttpServletResponse;
import javax.servlet.http.HttpSession;
public class LoginFilter implements Filter {
    private String logon_page;                              //登录页面
    private String logon_servlet;                           //登录 servlet 请求
    //消灭 filter 方法
    public void destroy() {
    }
    //过滤器服务方法
    public void doFilter(ServletRequest request, ServletResponse response,
            FilterChain chain) throws IOException, ServletException {
        HttpServletRequest req = (HttpServletRequest) request;
        HttpServletResponse resp = (HttpServletResponse) response;
        resp.setContentType("text/html;");
        resp.setCharacterEncoding("GBK");
        HttpSession session = req.getSession();
        PrintWriter out = resp.getWriter();
        // 得到用户请求的 URI
        String request_uri = req.getRequestURI();
        // 得到 Web 应用程序的上下文路径
        String ctxPath = req.getContextPath();
        // 去除上下文路径,得到剩余部分的路径
        String uri = request_uri.substring(ctxPath.length());
        // 判断用户访问的是否是登录页面或提交登录请求
        if (uri.equals(logon_page)||uri.equals(logon_servlet)) {
            //执行下一个过滤器
            chain.doFilter(request, response);
        } else {
            // 如果访问的不是登录页面,则判断用户是否已经登录
            if (null != session.getAttribute("user")
                    && "" != session.getAttribute("user")) {
                //执行下一个过滤器
                chain.doFilter(request, response);
            } else {
                out.println("您没有登录,请先登录!3 秒钟后回到登录页面.");
                resp.setHeader("refresh", "3;url=" + ctxPath + logon_page);
                return;
            }
        }
    }
    //过滤器初始化方法
    public void init(FilterConfig config) throws ServletException {
        // 从 web.xml 的部署描述符中获取登录页面
        logon_page = config.getInitParameter("login_uri"); //获得参数 login_uri 的值
        logon_servlet = config.getInitParameter("login_Servlet"); //获得参数 login_uri 的值
    }
}
```

代码模板 LoginServlet.java 如下：

```
package servlet;
import java.io.IOException;
import javax.servlet.ServletException;
import javax.servlet.http.HttpServlet;
import javax.servlet.http.HttpServletRequest;
import javax.servlet.http.HttpServletResponse;
import javax.servlet.http.HttpSession;
public class LoginServlet extends HttpServlet {
    protected void doGet(HttpServletRequest request,
            HttpServletResponse response) throws ServletException, IOException {
        String username=request.getParameter("name");
        String password=request.getParameter("pwd");
        if(username!=null&&username.equals("filter")){
            if(password!=null&&password.equals("filter")){
                HttpSession session=request.getSession();
                session.setAttribute("user", username);
                response.sendRedirect("loginSuccess.jsp");
            }else{
                response.sendRedirect("login.jsp");
            }
        }else{
            response.sendRedirect("login.jsp");
        }
    }
    protected void doPost(HttpServletRequest request,
            HttpServletResponse response) throws ServletException, IOException {
        doGet(request,response);
    }
}
```

代码模板 login.jsp 如下：

```
<%@ page language="java" contentType="text/html; charset=GBK" pageEncoding=
"GBK"%>
<html>
  <head>
    <title>login.jsp</title>
  </head>
  <body bgcolor="lightPink">
    <form action="isLogin" method="post">
        <table>
            <tr>
                <td>用户名：</td>
                <td><input type="text" name="name"/></td>
```

```
            </tr>
            <tr>
                <td>密 码: </td>
                <td><input type="password" name="pwd"/></td>
            </tr>
            <tr>
                <td><input type="submit" value="提交"/></td>
                <td><input type="reset" value="重置"/></td>
            </tr>
        </table>
    </form>
  </body>
</html>
```

代码模板 loginSuccess.jsp 如下：

```
<%@ page language="java" contentType="text/html; charset=GBK" pageEncoding=
"GBK"%>
<html>
<head>
<title>loginSuccess.jsp</title>
</head>
<body>
    <%
    String username=(String)session.getAttribute("user");
    %>
    恭喜<%=username %>登录成功!
</body>
</html>
```

上述任务中的过滤器，要首先检查用户请求的 URL 是不是 login.jsp 或者登录请求(isLogin)，这两个值都放在了过滤器的初始化参数中。如果用户访问的是 login.jsp 或者登录请求，过滤器就执行 chain..doFilter()继续请求。如果用户访问的不是 login.jsp 或者登录请求，则过滤器先判断用户是否登录成功，登录成功则执行 chain..doFilter()继续请求，否则重定向到 login.jsp。

9.3.3 拓展训练

1. 简述过滤器的运行原理。
2. Filter 接口中有哪些方法？它们分别具有哪些功能？
3. 在 web.xml 中部署过滤器需要哪些标记？这些标记的作用是什么？
4. 在任务的 Web 工程 loginValidate 中再新建几个 JSP 页面，在没有登录成功的情况下，运行这几个 JSP 页面看看是什么效果。

9.4 小　　结

- 在 JSP/Servlet 要实现过滤器，必须实现 Filter 接口，并在 web. xml 中定义部署过滤器，让服务器知道加载哪个过滤器。
- Filter 接口有 init()、doFilter()与 destroy()三个方法。这三个方法与 Servlet 接口的 init()、service()与 destroy()类似。
- 过滤器必须在 web. xml 中部署，可以使用<filter>和<filter-mapping>标记部署过滤器。其中使用<filter-name>部署过滤器的名称，使用<filter-class>部署过滤器的类名，使用<url-pattern>部署 URL 的请求模式。

第 10 章

表达式语言

在前面章节中编写 JSP 页面时,经常使用 Java 代码来实现页面显示逻辑。网页中夹杂着 HTML 与 Java 代码,给网站的设计与维护带来困难。我们可以使用表达式语言(Expression Language,EL)来访问和处理应用程序的数据。这样 JSP 页面就尽量减少了 Java 代码的使用,为以后的工作提供了方便。本章将重点介绍表达式语言 EL 的基本用法。

10.1 使用 EL 访问对象的属性

10.1.1 知识要点

EL 是 JSP 2.0 规范中增加的,它的基本语法为:

```
${表达式}
```

类似于 JSP 表达式<%=表达式%>,EL 语句中的表达式值会被直接送到浏览器显示。使用 EL 可以获取对象的属性,如 JavaBean、数组或 List 对象。

1. 获取 JavaBean 的属性值

假设在 JSP 页面中有这样一条语句:

```
<jsp:getProperty property="age" name="user"/>
```

那么,可以使用 EL 获取 user 的属性 age,修改如下:

```
${user.age}
```

其中,点运算符前面为 JavaBean 的对象 user,后面为该对象的属性 age,表示利用 user 对象的 getAge()方法取得值,而后显示在网页上。

2. 获取数组中的元素

假设在 JSP 页面中有这样一段代码:

```
<%
    String dogs[]={"lili","huahua","guoguo"};
    request.setAttribute("array", dogs);
%>
```

那么,在页面某处可以使用 EL 取出数组中的元素,代码如下:

```
${array[0]}
${array[1]}
${array[2]}
```

3. 获取 List 对象中的元素

假设在 JSP 页面中有这样一段代码：

```
<%
    ArrayList<UserBean> users=new ArrayList<UserBean>();
    UserBean ub1=new UserBean("zhang",20);
    UserBean ub2=new UserBean("zhao",50);
    users.add(ub1);
    users.add(ub2);
    request.setAttribute("array", users);
%>
```

其中，UserBean 有两个属性：name 和 age，那么在页面某处可以使用 EL 取出 UserBean 中的属性，代码如下：

```
${array[0].name} ${array[0].age}
${array[1].name} ${array[1].age}
```

10.1.2 技能操作

使用 EL 表达式取出对象的属性。具体任务如下：

编写一个创建 JavaBean 对象的类 UserBean，该类中有两个属性（成员变量）：name 和 age，然后在 JSP 页面 eg10_1.jsp 中使用 EL 取出该 JavaBean 对象的属性。

代码模板 UserBean.java 如下：

```
package bean;
public class UserBean {
    String name;
    int age;
    public UserBean(){
        name="EL 表达式";
        age=5;
    }
    public String getName() {
        return name;
    }
    public void setName(String name) {
        this.name = name;
    }
    public int getAge() {
        return age;
```

```
    }
    public void setAge(int age) {
        this.age = age;
    }
}
```

代码模板 eg10_1.jsp 如下：

```
<%@ page language="java" contentType="text/html; charset=GBK" pageEncoding=
"GBK"%>
<%@ page import="bean.*" %>
<html>
<head>
<title>EL 表达式</title>
</head>
<body>
    <jsp:useBean id="user" class="bean.UserBean" scope="page"/>
    姓名：${user.name} <!-- 使用 EL 取出 name 属性 -->
    <br>
    年龄：${user.age} <!-- 使用 EL 取出 age 属性 -->
</body>
</html>
```

代码分析：

• EL 处理 null 值。

对于 null 值直接以空字符串显示，而不是 null，运算时也不会发生错误或空指针异常。所以在使用 EL 访问对象的属性时，不需要判断对象是否为 null 对象。这样就为编写程序提供了方便。

• JSP 页面与 EL。

JSP 页面默认支持 EL，但如果 JSP 页面使用 page 指令设置 isELIgnored 属性(默认为 false)值为 true，则该页面不能使用 EL。

• 对象的有效范围。

在 EL 中，可以使用 EL 内置对象指定范围来访问属性，EL 内置对象将在稍后介绍。如果不指定对象的有效范围，则以 page、request、session、application 的顺序查找 EL 中所指定的对象。

• (.)与[]运算符的区别。

EL 中点运算符(.)和[]运算符，在一些情况下用法是一样的，总结如下：

(1) (.)运算符左边可以是 JavaBean 或 Map 对象。

(2) []运算符左边可以是 JavaBean、Map、数组或 List 对象。

使用 EL 如何取得 Map 对象中的值呢？假设在 JSP 页面中有这样一段代码：

```
<%
    HashMap<String, String> map=new HashMap<String, String>();
    map.put("fisrt", "第一");
    map.put("second", "第二");
    request.setAttribute("number", map);
%>
```

那么在页面某处可以使用 EL 获得 Map 中的值,代码如下:

```
${number.fisrt}
${number.second}
```

或

```
${number["fisrt"]}
${number["second"]}
```

10.1.3 拓展训练

如果把上述 eg10_1.jsp 中的代码:

```
<jsp:useBean id="user" class="bean.UserBean" scope="page"/>
```

删除,然后再运行程序,查看页面显示结果;如果不使用 EL 取值,而使用标记<jsp:getProperty>取值,那么再删除

```
<jsp:useBean id="user" class="bean.UserBean" scope="page"/>
```

的情况下,运行网页会是什么结果?

10.2 EL 内置对象

EL 内置对象共有 11 个,本节只是介绍几个常用的 EL 内置对象:pageScope、requestScope、sessionScope、applicationScope、param 以及 paramValues。

10.2.1 知识要点

1. 与作用范围相关的内置对象

与作用范围相关的 EL 内置对象有 pageScope、requestScope、sessionScope 和 applicationScope,分别可以获取 JSP 内置对象 pageContext、request、session 和 application 中的数据。如果在 EL 中没有使用内置对象指定作用范围,则从作用范围为 pageScope 的数据开始寻找。获取数据的格式如下:

```
${EL 内置对象.关键字对象.属性}
```

或

```
${EL 内置对象.关键字对象}
```

例如:

```
<jsp:useBean id="user" class="bean.UserBean" scope="page"/>
<jsp:setProperty name="user" property="name" value="EL 内置对象" />
name: ${pageScope.user.name}
```

再比如,在 JSP 页面中有这样一段代码:

```
<%
    ArrayList<UserBean> users=new ArrayList<UserBean>();
    UserBean ub1=new UserBean("zhang",20);
    UserBean ub2=new UserBean("zhao",50);
    users.add(ub1);
    users.add(ub2);
    request.setAttribute("array", users);
%>
```

其中,UserBean 有两个属性: name 和 age,那么在 request 有效的范围内可以使用 EL 取出 UserBean 的属性,代码如下:

```
${requestScope.array[0].name} ${requestScope.array[0].age}
${requestScope.array[1].name} ${requestScope.array[1].age}
```

2. 与请求参数相关的内置对象

与请求参数相关的 EL 内置对象有 param 和 paramValues。获取数据的格式如下:

```
${EL 内置对象.参数名}
```

例如,input.jsp 的代码如下:

```
<form method = "post" action = "param.jsp">
    <p>姓名: <input type="text" name="username" size="15" /></p>
    <p>兴趣:
    <input type="checkbox" name="habit" value="看书"/>看书
    <input type="checkbox" name="habit" value="玩游戏"/>玩游戏
    <input type="checkbox" name="habit" value="旅游"/>旅游
    <p>
    <input type="submit" value="提交"/>
</form>
```

那么,在 param.jsp 页面中可以使用 EL 获取参数值,代码如下:

```
<%request.setCharacterEncoding("GBK");%>
<body>
```

```
<h2>EL 隐含对象 param、paramValues</h2>
姓名：${param.username}</br>
兴趣：
${paramValues.habit[0]}
${paramValues.habit[1]}
${paramValues.habit[2]}
```

10.2.2 技能操作

使用 EL 内置对象从 JSP 内置对象中获取数据。具体任务如下：

编写一个 Servlet 类，在该类中使用 request 内置对象存储数据，然后从该 servlet 转发到 show.jsp 页面，最后在 show.jsp 页面中显示 request 内置对象的数据。首先，运行 servlet，在 IE 地址栏中输入：

```
http://localhost:8080/ch10/saveServlet
```

程序运行结果如图 10-1 所示。

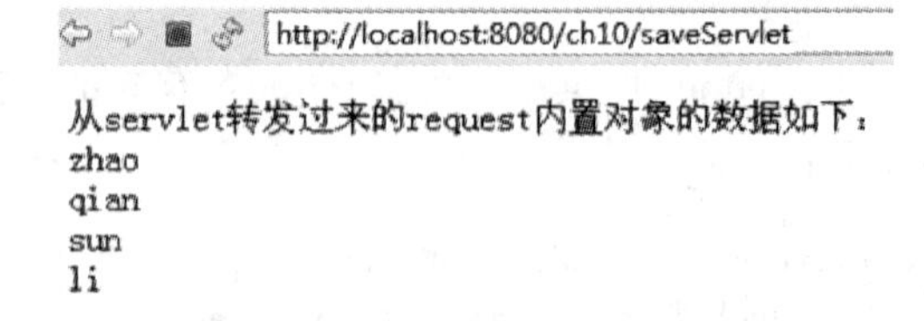

图 10-1 使用 EL 内置对象获取 JSP 内置对象的数据

配置文件 web.xml 如下：

```
<servlet>
    <servlet-name>saveServlet</servlet-name>
    <servlet-class>servlet.SaveServlet</servlet-class>
</servlet>
<servlet-mapping>
    <servlet-name>saveServlet</servlet-name>
    <url-pattern>/saveServlet</url-pattern>
</servlet-mapping>
```

代码模板 SaveServlet.java 如下：

```
package servlet;
import java.io.IOException;
import javax.servlet.RequestDispatcher;
import javax.servlet.ServletException;
import javax.servlet.http.HttpServlet;
import javax.servlet.http.HttpServletRequest;
```

```
import javax.servlet.http.HttpServletResponse;
public class SaveServlet extends HttpServlet {
    protected void doGet(HttpServletRequest request, HttpServletResponse response)
            throws ServletException, IOException {
        String names[]={"zhao","qian","sun","li"};
        request.setAttribute("name", names);
        RequestDispatcher dis=request.getRequestDispatcher("show.jsp");
        dis.forward(request, response);
    }
    protected void doPost(HttpServletRequest request, HttpServletResponse response)
            throws ServletException, IOException {
        doGet(request,response);
    }
}
```

代码模板 show.jsp 如下：

```
<%@ page language="java" contentType="text/html; charset=GBK" pageEncoding=
"GBK"%>
<html>
<head>
    <title>EL 内置对象</title>
</head>
<body>
    从 servlet 转发过来的 request 内置对象的数据如下：<br>
${requestScope.name[0]}<br><!--使用 EL 内置对象 requestScope 取出 request 中数组的第 1 个
元素-->
${requestScope.name[1]}<br><!--使用 EL 内置对象 requestScope 取出 request 中数组的第 2 个
元素-->
${requestScope.name[2]}<br><!--使用 EL 内置对象 requestScope 取出 request 中数组的第 3 个
元素-->
${requestScope.name[3]}<br><!--使用 EL 内置对象 requestScope 取出 request 中数组的第 4 个
元素-->
</body>
</html>
```

通过上面的例子可以看出，EL 内置对象与 JSP 内置对象不同，EL 内置对象仅仅代表作用范围。

10.2.3 拓展训练

1. 在 Web 应用程序中有以下程序代码段，执行后转发到某个 JSP 页面：

```
ArrayList<String> dogNames=new ArrayList<String>();
dogNames.add("goodDog");
request.setAttribute("dogs", dogNames);
```

以下(　　)选项可以正确地使用 EL 取得数组中的值。

A. ${ dogs .0}　　B. ${ dogs [0]}

C. ${ dogs .[0]}　　D. ${ dogs "0"}

2. (　　)不是 EL 的内置对象。

A. request　　B. pageScope

C. sessionScope　　D. applicationScope

3. 如果把任务中 show. jsp 页面中的代码:

```
${requestScope.name[0]}<br>
${requestScope.name[1]}<br>
${requestScope.name[2]}<br>
${requestScope.name[3]}<br>
```

改成:

```
${name[0]}<br>
${name[1]}<br>
${name[2]}<br>
${name[3]}<br>
```

然后运行程序,查看运行结果。

10.3 小　　结

- 在 JSP 页面中一些简单的属性、请求参数等值的获取,一些简单的运算或判断,可以使用 EL 表达式来处理,减少了页面中的 Java 代码。
- 可以使用 EL 表达式获取对象的属性值,例如 JavaBean 对象、Map 对象、数组或 List 对象,还可以使用它获取 JSP 内置对象中的数据。

第 11 章

标准标签库

在网站开发与制作过程中，可以使用标准标签库(JavaServer Pages Standard Tag Library，JSTL)来替换网页中实现页面显示逻辑的 Java 代码。这样也可以减少 JSP 页面中 Java 代码的使用，为后续的维护工作提供方便。本章将重点介绍标准标签库 JSTL 的基本用法。

11.1 一般用途的标签

JSTL 是一个标准规范，但不在 JSP 的规范中，所以需要下载 JSTL 实现(jar 包)。可以登录网站

https://jstl.dev.java.net/

下载 JSTL1.2 的 jar 包：jstl-impl-1.2.jar。另外，还需要 JSTL 标准接口与类(jstl.jar)。

如果使用 Tomcat 作为 Web 服务器，可以在 Tomcat 的 webapps\examples\WEB-INF\lib 中找到 jstl.jar 文件。在 JSP 页面中要想使用 JSTL 核心标签库，必须把 jstl-impl-1.2.jar 与 jstl.jar 复制到 Web 工程的 WEB-INF\lib 中。同时在 JSP 页面中使用 taglib 标记定义前置名称与 uri 引用，代码如下：

```
<%@ taglib prefix="c" uri="http://java.sun.com/jsp/jstl/core"%>
```

本书中只说明 JSTL 核心标签库中几个常用的标签，其他标签请参考 JSTL 说明文档或专门的书籍。

11.1.1 知识要点

1. <c:out>标签

<c:out>用来显示数据的内容，与<%=表达式%>或 ${表达式}类似。格式如下：

```
<c:out value="输出的内容" [default="defaultValue"]/>
```

或

```
<c:out value="输出的内容">
    defaultValue
</c:out>
```

其中，value 值可以是一个 EL 表达式，也可以是一个字符串；default 可有可无，当 value 值不存在时，就输出 defaultValue。例如：

```
<c:out value="${param.data}" default="No Data" />
<br>
<c:out value="${param.nothing}" />
<br>
<c:out value="This is a String" />
```

输出的结果如图 11-1 所示。

http://localhost:8080/ch11/NewFile.jsp

No Data

This is a String

图 11-1　<c:out>标签

2. <c:set>标签

• 设置作用域变量。

可以使用<c:set>在 page、request、session、application 等范围内设置一个变量。一般格式如下：

```
<c:set value="value" var="varName" [scope="page|request|session|application"]/>
```

将 value 值赋值给变量 varName。例如：

```
<c:set value="zhao" var="userName" scope="session"/>
```

相当于

```
<% session.setAttribute("userName","zhao"); %>
```

• 设置 JavaBean 的属性。

使用<c:set>设置 JavaBean 的属性时，必须使用 target 属性进行设置。其格式如下：

```
<c:set value="value" target="target" property="propertyName"/>
```

将 value 赋值给 target 对象(JaveBean 对象)的 propertyName 属性。如果 target 为 null 或没有 set 方法则抛出异常。

3. <c:remove>标签

如果要删除某个变量，则可以使用<c:remove>标签。例如：

```
<c:remove var="userName" scope="session"/>
```

相当于

```
<%session.removeAttribute("userName") %>
```

11.1.2　技能操作

灵活使用 JSTL 基本输入输出标签，具体任务如下：

编写一个 JSP 页面 input_out.jsp，在该页面中使用<c:set>标签定义几个变量，并使用<c:out>标签输出这几个变量的值。运行效果如图 11-2 所示。

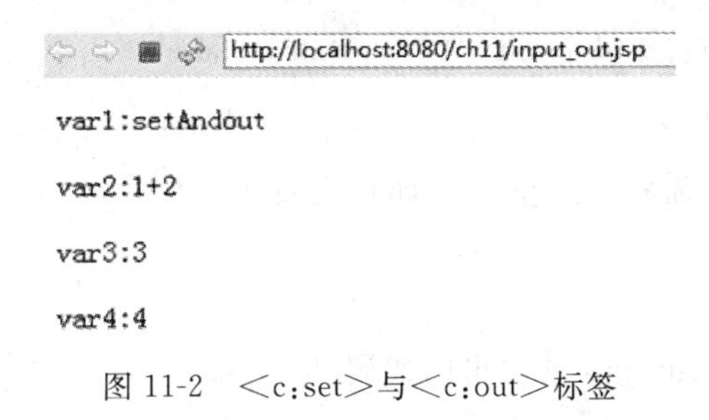

图 11-2　<c:set>与<c:out>标签

代码模板 input_out.jsp 如下：

```
<%@ page language="java" contentType="text/html; charset=GBK"
pageEncoding="GBK"%>
<%@ taglib prefix="c" uri="http://java.sun.com/jsp/jstl/core"%>
<html>
<head>
<title>input_out.jsp</title>
</head>
<body>
<c:set var="var1" value="setAndout" />
<c:set var="var2" value="1+2"/>
<c:set var="var3" value="${1+2}" />
<c:set var="var4 scope ="request" value="${1 + 2}" />
var1:<c:out value="${var1}" default="No Data" /><p>
var2:<c:out value="${var2}" default="No Data" /><p>
var3:<c:out value="${var3}" default="No Data" /><p>
var4:<c:out value="${var4+1}" default="No Data" /><p>
</body>
</html>
```

在 10.2 节中使用 EL 表达式就可以输出变量或表达式的值，那么我们为什么还要学习<c:out>标签呢？下面先来猜猜这段程序的输出结果是什么？

```
<%
    String s="<p>有特殊字符</p>";
    request.setAttribute("exp", s);
%>
${exp}
```

运行时才发现其中的 HTML 标记<p>没有起到创建段落的作用。如果希望<p>达到创建段落的作用，那么必须把语句：

```
${exp}
```

改成：

```
<c:out value="<p>有特殊字符</p>" escapeXml="false" />
```

默认情况下，<c:out>将<、>、'、"和 & 转换为 <、>、'、"和&，如果不想转换，只需将 escapeXml 属性设置为 false。

11.1.3 拓展训练

把前述任务中 input_out.jsp 页面里的代码：

```
var1:<c:out value="${var1}" default="No Data" /><p>
var2:<c:out value="${var2}" default="No Data" /><p>
var3:<c:out value="${var3}" default="No Data" /><p>
var4:<c:out value="${var4+1}" default="No Data" /><p>
```

改成：

```
var1: ${var1}<p>
var2: ${var2}<p>
var3: ${var3}<p>
var4: ${var4+1}<p>
```

然后再运行程序，查看结果有什么不同。

11.2 条件控制标签

11.2.1 知识要点

1. <c:if>标签

<c:if>标签实现 if 语句的作用，具体语法格式如下：

```
<c:if test="条件表达式">
    主体内容
</c:if>
```

其中，条件表达式可以是 EL 表达式，也可以是 JSP 表达式。如果表达式的值为 true，则会执行<c:if>的主体内容，但是没有相对应的<c:else>标签。如果想在条件成立时执行一块内容，不成立时执行另一块内容，则可以使用<c:choose>、<c:when>及<c:otherwise>标签。

2. <c:choose>、<c:when>及<c:otherwise>标签

<c:choose>、<c:when>及<c:otherwise>标签实现 if/else if/else 语句的作用。具体语法格式如下：

```
<c:choose>
    <c:when test="条件表达式 1">
        主体内容 1
    </c:when>
    <c:when test="条件表达式 2">
        主体内容 2
    </c:when>
    <c:otherwise>
        表达式都不正确时,执行的主体内容
    </c:otherwise>
</c:choose>
```

11.2.2 技能操作

掌握<c:if>、<c:choose>、<c:when>及<c:otherwise>标签的使用方法。具体任务如下:

编写一个 JSP 页面 ifelse.jsp,在该页面中使用<c:set>标签把两个字符串设置在 request 内置对象中。使用<c:if>标签求出这两个字符串的最大值(按字典顺序比较大小),使用<c:choose>、<c:when>及<c:otherwise>标签求出这两个字符串的最小值。

代码模板 ifelse.jsp 如下:

```
<%@ page language="java" contentType="text/html; charset=GBK"
pageEncoding="GBK"%>
<%@ taglib prefix="c" uri="http://java.sun.com/jsp/jstl/core"%>
<html>
<head>
<meta http-equiv="Content-Type" content="text/html; charset=ISO-8859-1">
<title>ifelse.jsp</title>
</head>
<body>
<c:set value="if" var="firstNumber" scope="request"/>
<c:set value="else" var="secondNumber" scope="request"/>
<c:if test="${firstNumber>secondNumber}">
    最大值为${firstNumber}
</c:if>
<c:if test="${firstNumber<secondNumber}">
    最大值为${secondNumber}
</c:if>
<c:choose>
    <c:when test="${firstNumber<secondNumber}">
        最小值为${firstNumber}
    </c:when>
    <c:otherwise>
        最小值为${secondNumber}
```

```
        </c:otherwise>
    </c:choose>
    </body>
    </html>
```

为了更好地理解本例中的代码，将注意事项说明如下。

<c:when>及<c:otherwise>必须放在<c:choose>之中。当<c:when>的 test 结果为 true 时，会输出<c:when>的主体内容，而不理会<c:otherwise>的内容。<c:choose>中可有多个<c:when>，程序会从上到下进行条件判断，如果有个<c:when>的 test 结果为 true，就输出其主体内容，之后的<c:when>就不再执行。如果所有的<c:when>的 test 结果都为 false，则会输出<c:otherwise>的内容。<c:if>与<c:choose>也可以嵌套使用，例如：

```
<c:set value="fda" var="firstNumber" scope="request"/>
<c:set value="else" var="secondNumber" scope="request"/>
<c:set value="ddd" var="threeNumber" scope="request"/>

<c:if test="${firstNumber>secondNumber}">
    <c:choose>
        <c:when test="${firstNumber<threeNumber}">
            最大值为${threeNumber}
        </c:when>
        <c:otherwise>
            最大值为${firstNumber}
        </c:otherwise>
    </c:choose>
</c:if>

<c:if test="${secondNumber>firstNumber}">
    <c:choose>
        <c:when test="${secondNumber<threeNumber}">
            最大值为${threeNumber}
        </c:when>
        <c:otherwise>
            最大值为${secondNumber}
        </c:otherwise>
    </c:choose>
</c:if>
```

11.2.3 拓展训练

编写一个 JSP 页面 useTag.jsp，在该页面中使用<c:set>标签把三个字符串设置在 request 内置对象中。使用<c:if>、<c:when>、<c:otherwise>和<c:choose>标签求出

这三个字符串的最小值。

11.3　迭 代 标 签

11.3.1　知识要点

JSTL 的<c:forEach>标签可以实现程序中的 for 循环，完成迭代功能。语法格式如下：

```
<c:forEach var="变量名" items="数组或 Collection 对象">
    循环体
</c:forEach>
```

其中，items 属性可以是数组或 Collection 对象，每次循环读取对象中的一个元素，并赋值给 var 属性指定的变量，之后就可以在循环体使用 var 指定的变量获取对象的元素。例如，在 JSP 页面中有这样一段代码：

```
<%
    ArrayList<UserBean> users=new ArrayList<UserBean>();
    UserBean ub1=new UserBean("zhao",20);
    UserBean ub2=new UserBean("qian",40);
    UserBean ub3=new UserBean("sun",60);
    UserBean ub4=new UserBean("li",80);
    users.add(ub1);
    users.add(ub2);
    users.add(ub3);
    users.add(ub4);
    request.setAttribute("usersKey", users);
%>
```

那么，在页面某处可以使用<c:forEach>循环遍历出数组中的元素。代码如下：

```
<table>
  <tr>
    <th>姓名</th>
    <th>年龄</th>
  </tr>
<c:forEach var="user" items="${requestScope.usersKey}">
  <tr>
    <td>${user.name}</td>
    <td>${user.age}</td>
  </tr>
</c:forEach>
</table>
```

11.3.2 技能操作

使用<c:forEach>标签循环遍历集合中的元素。具体任务如下：

把8.2.2节中showAllGoods.jsp页面里的for语句改成<c:forEach>标签。

代码模板showAllGoods.jsp如下：

```
<%@ page language="java" contentType="text/html; charset=GBK" pageEncoding=
"GBK"%>
<%@ taglib prefix="c" uri="http://java.sun.com/jsp/jstl/core"%>
<%@ page import="java.util.ArrayList" %>
<%@ page import="bean.Goods" %>
<html>
<head>
<title>showAllGoods.jsp</title>
</head>
<body>
    <table border="1">
        <tr>
            <th>商品编号</th>
            <th>商品名称</th>
            <th>商品价格</th>
            <th>商品类别</th>
        </tr>
    <c:forEach var="good" items="${requestScope.goods}">
        <tr>
            <td>${good.goodsId}</td>
            <td>${good.goodsName}</td>
            <td>${good.goodsPrice}</td>
            <td>${good.goodsType}</td>
        </tr>
    </c:forEach>
    </table>
</body>
</html>
```

说明：在有些情况下，我们需要为<c:forEach>标签里的var属性指定初始值(begin)、结束值(end)和步长(step)。例如，下面的testTag.jsp文件：

```
<table border=1>
   <tr>
      <th>number</th><th>Square</th>
   </tr>
<c:forEach var="x" begin="1" end="5" step="1">
   <tr>
      <td><c:out value="${x}"/></td>
```

```
        <td><c:out value="${x + x}"/></td>
    </tr>
</c:forEach>
</table>
```

该程序运行结果如图 11-3 所示。

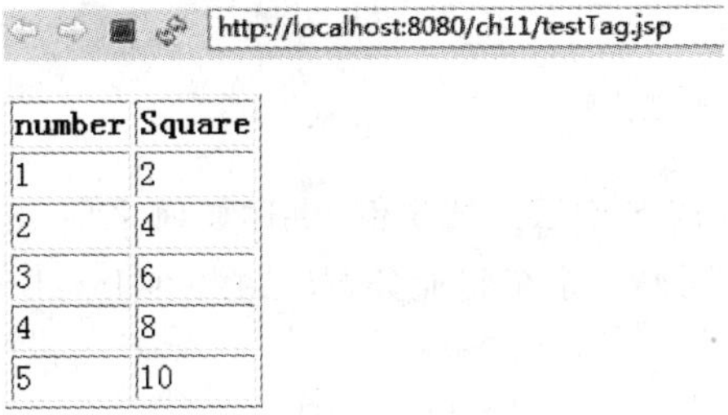

http://localhost:8080/ch11/testTag.jsp

number	Square
1	2
2	4
3	6
4	8
5	10

图 11-3　<c:forEach>标签

11.3.3　拓展训练

1. (　　)JSTL 标签可以实现 Java 程序中的 if 语句功能。

   A. <c:set>　　B. <c:out>　　C. <c:forEach>　　D. <c:if>

2. (　　)JSTL 标签可以实现 Java 程序中的 for 语句功能。

   A. <c:set>　　B. <c:out>　　C. <c:forEach>　　D. <c:if>

3. 编写一个 JSP 页面 mulTable.jsp，在该页面中使用<c:forEach>标签输出九九乘法表。

11.4　小　　结

在 JSP 页面中可以使用 JSTL 标签来替代实现页面逻辑的 Java 程序。例如，<c:if>替代 if 语句和<c:forEach>替代 for 语句。

第12章

动态网站开发综合实例

本章通过一个小型的BBS论坛系统实例，讲述如何采用JSP+JavaBean+Servlet的模式来开发一个简单的动态网站。系统将业务逻辑封装在JavaBean中，使系统的可维护性和可扩展性大大提高。

系统的开发环境如下：

- 操作系统 Windows 7；
- 数据库 MySQL 5.5；
- JSP引擎 Tomcat 6.0；
- 集成开发环境(IDE)Eclipse IDE for JavaEE Developers。

12.1 系统分析与设计

12.1.1 系统需求分析

BBS(Bulletin Board System，电子公告牌系统)俗称论坛系统，是互联网上一种交互性极强、网友喜闻乐见的信息服务形式。根据相应的权限，论坛用户可以进行浏览信息、发布信息、回复信息、管理信息等操作，从而加强不同用户间的文化交流和思想沟通。

"BBS论坛系统"的主要功能有用户管理、版块管理、帖子管理、友情链接管理、广告管理。

其中各主要功能大体说明如下：

(1) 用户管理主要包括用户注册、用户登录、用户资料修改等功能。

(2) 版块管理主要包括增加版块、编辑版块、删除版块等功能。

(3) 帖子管理主要包括发布帖子、回复帖子、搜索帖子、浏览帖子、转移帖子、编辑帖子、删除帖子、帖子加精、帖子置顶等功能。

(4) 友情链接管理主要包括增加链接、修改链接、删除链接等功能。

(5) 广告管理主要包括放置广告、删除广告等功能。

另外，需要说明的是，以上各项功能中有些只需要普通用户权限就能够完成，而有些则需要版主或管理员权限才能完成。

结合前述需求分析，可以给出论坛系统的总体结构图，如图12-1所示。

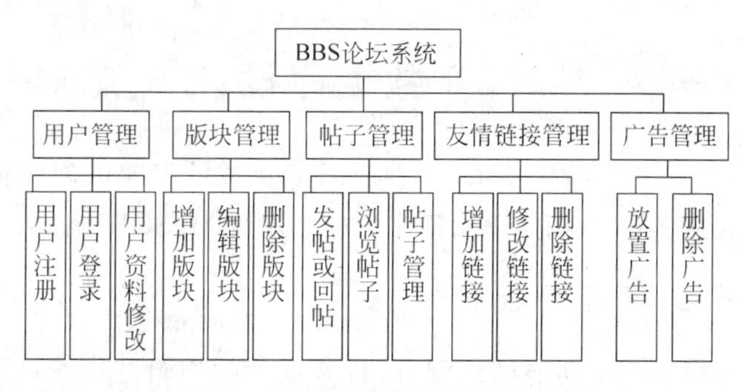

图 12-1　"BBS 论坛系统"总体结构图

12.1.2　系统功能模块划分

注册用户可以根据权限使用 BBS 论坛系统进行相应的操作，具体的系统功能模块如表 12-1 所示。

表 12-1　系统功能模块划分

模块名称	功能描述	输入内容	输出内容
用户注册	检测注册信息	用户名等注册信息	注册结果(是否成功)
用户登录	合法用户通过验证进入系统	用户名、密码	登录状态(是否成功)
用户资料修改	根据当前状况修改个人信息	修改的信息	修改信息(是否成功)
发布帖子	根据需要发布帖子	用户的言论	用户的言论
回复帖子	回复已存在的话题帖子	用户的回复	用户的回复
搜索帖子	根据需要搜索帖子	搜索条件	符合搜索条件的帖子
浏览帖子	浏览任意版块帖子	选择任意话题帖子	该话题帖子及其回复
转移帖子	根据实际情况移动帖子位置	"移动"命令	移动结果(是否成功)
编辑帖子	修改曾经发过的帖子	修改的内容	修改后的内容
删除帖子	删除违规帖子	"删除"命令	删除结果(是否成功)
帖子加精	将重要话题帖子列为精华帖子	"加精"命令	添加加精图标的帖子
帖子置顶	将重要话题帖子放置于最上方	"置顶"命令	添加置顶图标的帖子
添加版块	添加版块、设置版主	版块的相关信息	版块列表
修改版块	修改版块信息	版块的修改信息	修改结果(是否成功)
删除版块	删除版块	"删除"命令	删除结果(是否成功)
增加链接	接受友情链接申请，等待验证	友情网站的信息	友情网站的链接
修改链接	验证并修改友情链接信息	友情链接信息	修改后的友情链接
删除连接	清理不合格的友情链接	"删除"命令	删除结果(是否成功)
设置广告	选择已有位置发布广告	广告语、URL 地址	前台广告
删除广告	清理已发布的广告	"删除"命令	原有的广告消失

12.2 数据库设计

本系统利用 MySQL 5.5 创建 icefish 数据库，并在该数据库中创建相应的数据表；在具体实现上采用加载纯 Java 数据库驱动程序的方式连接 MySQL 5.5 数据库。

12.2.1 数据库逻辑结构设计

根据系统需求分析和系统功能模块划分，将数据库的逻辑结构设计为 MySQL 5.5 支持的实际数据模型，如表 12-2 至表 12-12 所示。

表 12-2 用户管理表

字 段	含 义	类型及长度	约 束
user_id	用户 ID	int(11)	auto_increment primary key
user_name	用户名	char(50)	
user_password	密码	char(100)	
user_sex	性别	char(2)	
user_birthday	生日	datetime	
user_QQ	QQ 号码	int(11)	
user_Email	电子邮件	char(50)	
user_tel	电话号码	char(50)	
user_face	头像	char(100)	
user_sign	个性签名	Text	
user_grade	用户等级	char(50)	
user_mark	用户头衔	int(11)	
user_topic	主题数	int(11)	
user_wealth	财富积分	int(11)	
user_post	帖子数	int(11)	
user_group	门派	char(50)	
user_lastip	上次登录 IP	char(15)	
user_delnum	删除数	int(11)	
user_friends	好友	text	
user_regtime	注册时间	datetime	
user_lasttime	最近登录时间	datetime	
user_locked	用户是否被锁定	ENUM("false","true")	default 'false'
user_admin	用户是否为后台管理员	ENUM("false","true")	default 'false'
user_password_a	密码提示问题答案	char(100)	
user_password_q	密码提示问题	char(100)	
user_age	年龄	int(11)	

续表

字　段	含　义	类型及长度	约　束
user_secondname	姓氏	char(50)	
user_truename	真实名字	char(50)	
user_blood	血型	char(10)	
user_shengxiao	生肖	char(10)	
user_nation	国家	char(50)	
user_province	省份	char(50)	
user_city	城市	char(50)	

表 12-3　用户信箱表

字　段	含　义	类型及长度	约　束
msg_id	消息 ID	int(11)	not null auto_increment primary key
msg_from	消息来源	char(50)	default null
msg_to	消息目的地	char(50)	not null
msg_topic	消息主题	char(100)	not null
msg_words	消息内容	mediumtext	not null
msg_time	消息传送时间	datetime	not null
msg_new	是否为新消息	enum('true','false')	not null default 'true'
msg_belong	待发消息	enum('true','false')	not null default 'true'
msg_del	删除消息	tinyint(1)	not null default '0'
msg_del_msgbin	接收消息	enum('0','1')	not null default '1'
msg_del_forebin	发送消息	enum('0','1')	not null default '1'

表 12-4　管理员表

字　段	含　义	类型及长度	约　束
admin_id	管理员 ID	int(11)	auto_increment primary key
admin_name	管理员用户名	char(50)	
admin_password	管理员密码	char(25)	
admin_user	普通用户名	char(50)	

表 12-5　帮助表

字　段	含　义	类型及长度	约　束
help_id	帮助 ID	int(11)	auto_increment primary key
help_name	帮助主题	char(100)	default null
help_info	帮助内容	text	

表 12-6　好友表

字　段	含　义	类型及长度	约　束
user_id	用户 ID	int(11)	default null
friend_id	好友 ID	int(11)	default null
friend_name	好友用户名	char(50)	default null
friend_addtime	添加好友时间	varchar(20)	default null
friend_Email	好友电子邮件	varchar(20)	default null

表 12-7　论坛版块管理表

字　段	含　义	类型及长度	约　束
board_id	版块 ID	int(11)	auto_increment primary key
board_isMother	是否为主版块	ENUM('true','false')	default 'false'
board_bid	所属版块 ID	int(11)	
board_name	版块名称	char(50)	
board_info	版块信息	text	
board_master	版主	char(50)	
board_img	版块图标	char(100)	
board_postnum	版块帖子数	int(11)	
board_topicnum	版块主题数	int(11)	
board_todaynum	版块今日帖数	int(11)	
board_lastreply	最新回复帖 ID	int(11)	

表 12-8　帖子管理表

字　段	含　义	类型及长度	约　束
post_id	帖子 ID	int(11)	auto_increment primary key
post_boardid	帖子所属版块	int(11)	
post_user	发帖用户	char(50)	
post_topicid	帖子主题 ID	int(11)	
post_replyid	帖子回复 ID	int(11)	

续表

字　段	含　义	类型及长度	约　束
post_content	帖子内容	text	
post_time	发帖时间	datetime	
post_edittime	帖子编辑时间	datetime	
post_ip	发帖 IP	char(15)	

表 12-9　论坛基本设置表

字　段	含　义	类型及长度	约　束
setting_id	设置 ID	int(11)	auto_increment primary key
setting_copyrights	版权设置	text	
setting_author	设置人	text	
setting_open	论坛是否处于开放状态,false 为论坛关闭,true 为开放	ENUM('false','true')	default 'true'
setting_offwords	舍入字设置	text	
setting_badwords	污秽字设置	text	
setting_bbsname	BBS 名称	char(100)	
setting_bbsurl	BBS 访问地址	char(100)	
setting_bbslogo	BBS 图标	char(100)	
setting_info	信息内容	text	
setting_maxonlinetime	最长在线时间	datetime	
setting_maxonline	最大在线数	int(11)	
setting_todaynum	今日访问量	int(11)	
setting_yesterdaynum	昨日访问量	int(11)	
setting_maxnum	最大访问量	int(11)	
setting_totaltopic	总主题数	int(11)	
setting_totalpost	总帖子数	int(11)	
setting_totaluser	总用户数	int(11)	

表 12-10　话题管理表

字　段	含　义	类型及长度	约　束
topic_id	话题 ID	int(11)	auto_increment primary key
topic_boardid	话题所属版块	int(11)	
topic_user	发布话题用户	char(50)	
topic_name	话题名称	char(100)	
topic_time	发布话题时间	datetime	

续表

字　段	含　义	类型及长度	约　束
topic_hits	话题点击量	int(11)	
topic_replynum	话题回复数	int(11)	
topic_lastreplyid	最新回复话题 ID	int(11)	
topic_top	是否为置顶帖	ENUM("false","true")	default 'false'
topic_best	是否为精华	ENUM("false","true")	default 'false'
topic_del	是否为删除帖	ENUM("false","true")	default 'false'
topic_hot	是否为热门帖	ENUM("false","true")	default 'false'

表 12-11　友情链接表

字　段	含　义	类型及长度	约　束
link_id	链接 ID	int(11)	auto_increment primary key
link_name	链接名称	char(50)	
link_url	链接地址	char(100)	
link_info	链接信息	text	
link_logo	链接图标	char(100)	
link_islogo	链接是否为图标	ENUM('true','false')	not null default 'false'
link_pass	链接过审	ENUM('true','false')	not null default 'false'

表 12-12　广告表

字　段	含　义	类型及长度	约　束
ad_id	广告 ID	int(11)	not null
ad_img	广告图片	char(100)	null
ad_url	广告链接地址	char(100)	null
ad_title	广告标题	char(100)	null

12.2.2　创建数据库和数据表

根据数据库的逻辑结构，创建数据库和创建数据表的代码如下：

```
/* 创建数据库 icefish */
drop database if exists icefish;
CREATE DATABASE icefish DEFAULT CHARACTER SET gb2312;
use icefish;

/* 创建用户管理表 icefish_user */
```

```
CREATE TABLE icefish_user (
  user_id int(11) auto_increment PRIMARY KEY
  user_name char(50) ,
  user_password char(100),
  user_sex char(2),
  user_birthda datetime,
  user_QQ int(11),
  user_Email char(50),
  user_tel char(50),
  user_face char(100),
  user_sign text,
  user_grade char(50),
  user_mark int(11),
  user_topic int(11),
  user_wealth int(11),
  user_post int(11),
  user_group char(50),
  user_lastip char(15),
  user_delnum int(11),
  user_friends text,
  user_regtime datetime,
  user_lasttime datetime,
  user_locked enum("false","true") default 'false',
  user_admin enum("false","true") default 'false',
  user_password_a char(100),
  user_password_q char(100),
  user_age int(11),
  user_secondname char(50),
  user_truename char(50),
  user_blood char(10),
  user_shengxiao char(10),
  user_nation char(50),
  user_province char(50),
  user_city char(50)
) ENGINE=InnoDB DEFAULT CHARSET=gb2312;

/*创建用户信箱表 icefish_msg */
CREATE TABLE icefish_msg (
  msg_id int(11) not null auto_increment,
  msg_from char(50) default null,
  msg_to char(50) not null,
  msg_topic char(100) not null,
  msg_words mediumtext not null,
  msg_time datetime not null,
  msg_new enum('true','false') not null default 'true',
```

```
    msg_belong enum('true','false') not null default 'true',
    msg_del tinyint(1) not null default '0',
    msg_del_msgbin enum('0','1') not null default '1',
    msg_del_forebin enum('0','1') not null default '1',
    PRIMARY KEY (msg_id)
) ENGINE=InnoDB DEFAULT CHARSET=gb2312;

/* 创建管理员表 icefish_admin */
CREATE TABLE icefish_admin (
    admin_id int(11) auto_increment PRIMARY KEY,
    admin_name char(50) ,
    admin_password char(25) ,
    admin_user char(50)
) ENGINE=InnoDB DEFAULT CHARSET=gb2312;

/* 创建帮助表 icefish_help */
CREATE TABLE icefish_help (
    help_id int(11) auto_increment PRIMARY KEY,
    help_name char(100) default null,
    help_info text
    ) ENGINE=InnoDB DEFAULT CHARSET=gb2312;

/* 创建好友表 icefish_friend */
CREATE TABLE icefish_friend (
    user_id int(11) default null,
    friend_id int(11) default null,
    friend_name char(50) default null,
    friend_addtime varchar(20) default null,
    friend_email varchar(20) default null
) ENGINE=InnoDB DEFAULT CHARSET=gb2312;

/* 创建论坛版块管理表 icefish_board */
CREATE TABLE icefish_board (
    board_id int(11) auto_increment PRIMARY KEY,
    board_isMother enum('true','false') default 'false',            /* 是否主版块 */
    board_bid int(11) ,
    board_name char(50) ,
    board_info text ,
    board_master char(50),
    board_img char(100),
    board_postnum int(11),
    board_topicnum int(11),
    board_todaynum int(11),
    board_lastreply int(11) /* 最新回复帖的 ID */
) ENGINE=InnoDB DEFAULT CHARSET=gb2312;
```

```
/*创建帖子管理表 icefish_post*/
CREATE TABLE icefish_post (
  post_id int(11) auto_increment PRIMARY KEY,
  post_boardid int(11) ,
  post_user char(50) ,
  post_topicid int(11),
  post_replyid int(11),
  post_content text,
  post_time datetime ,
  post_edittime datetime,
  post_ip char(15)
) ENGINE=InnoDB DEFAULT CHARSET=gb2312;

/*创建论坛基本设置表 icefish_setting*/
CREATE TABLE icefish_setting (
  setting_id int(11) auto_increment PRIMARY KEY,
  setting_copyrights text,
  setting_author text,
  setting_open enum('false','true') default 'true',
  /*论坛是否处于开放状态,false 为论坛关闭,true 为开放*/
  setting_offwords text,
  setting_badwords text,
  setting_bbsname char(100),
  setting_bbsurl char(100),
  setting_bbslogo char(100),
  setting_info text,
  setting_maxonlinetime datetime,
  setting_maxonline int(11),
  setting_todaynum int(11),
  setting_yesterdaynum int(11),
  setting_maxnum int(11),
  setting_totaltopic int(11),
  setting_totalpost int(11),
  setting_totaluser int(11)
) ENGINE=InnoDB DEFAULT CHARSET=gb2312;

/*创建话题管理表 icefish_topic*/
CREATE TABLE icefish_topic (
  topic_id int(11) auto_increment PRIMARY KEY,
  topic_boardid int(11) ,
  topic_user char(50) ,
  topic_name char(100) ,
  topic_time datetime ,
  topic_hits int(11),
  topic_replynum int(11),
```

```
  topic_lastreplyid int(11),
  topic_top enum("false","true") default 'false',              /*是否为置顶帖*/
  topic_best enum("false","true") default 'false',             /*是否为精华帖*/
  topic_del enum("false","true") default 'false',              /*是否为删除帖*/
  topic_hot enum("false","true") default 'false'               /*是否为热门帖*/
) ENGINE=InnoDB DEFAULT CHARSET=gb2312;

/*创建友情链接表 icefish_link*/
CREATE TABLE icefish_link (
  link_id int(11) auto_increment PRIMARY KEY,
  link_name char(50) ,
  link_url char(100) ,
  link_info text,
  link_logo char(100),
  link_islogo enum('true','false') not null default 'false',
  link_pass enum('true','false') not null default 'false'
) ENGINE=InnoDB DEFAULT CHARSET=gb2312;

/*创建广告表 icefish_ad*/
CREATE TABLE icefish_ad (
  ad_id int(11) not null ,
  ad_img char(100) null,
  ad_url char(100) null,
  ad_title char(100) null
) ENGINE=InnoDB DEFAULT CHARSET=gb2312;
```

12.3 系统管理

12.3.1 导入相关的Jar包

在本章中新建一个Web工程icefish,在icefish中开发本系统。由于本系统采用纯Java数据库驱动程序连接MySQL 5.5数据库,所以需要把Java数据库驱动程序mysql-connector-java-5.0.6.jar复制到icefish/WEB-INF/lib(本系统虚拟Web服务目录)及D:\Tomcat 6.0\common\lib(Web服务器安装目录)中。需要注意的是,如果在后期编码调试过程中出现"bad handshake error!"之类的错误提示,则需要把其他版本的Java数据库驱动程序删除。

12.3.2 JSP页面管理

本论坛系统中用到的JSP页面保存在icefish工程中,系统中有些JSP页面使用到CSS和JavaScript,有关CSS与JavaScript的知识,在本书第1章中有简单介绍,如果想深入理解和掌握这两方面内容可参考查阅相关书籍。

BBS 论坛系统首页为 index.jsp，后续操作都以此为基础展开，在浏览器地址栏中输入 http://localhost:8080/icefish 即可进入首页，运行效果如图 12-2 所示。

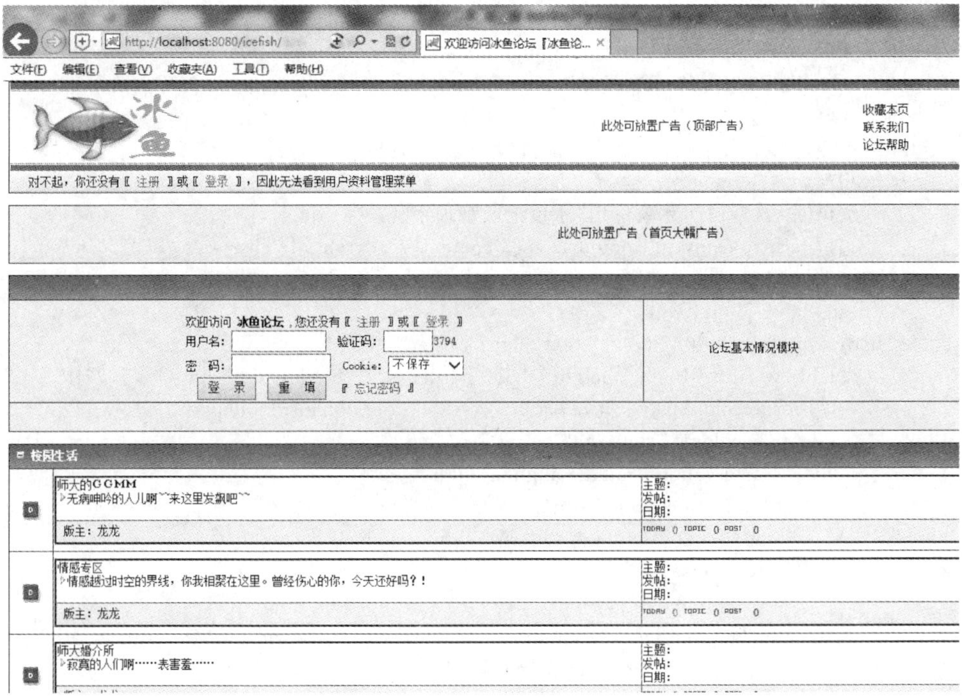

图 12-2　BBS 论坛系统首页效果

例子代码 index.jsp 如下：

```
<%@page pageEncoding="gb2312"%>
<%@page contentType="text/html;charset=gb2312" %>
<%@page language="java"%>
<%@page import="java.sql.*"%>
<%@page import="java.util.*" %>
<%@include file="top.jsp" %>
<%
String sql1="select * from icefish_board WHERE board_isMother='true' order by board_id asc";
ResultSet rs1=null;
rs1=indexBean.executeSQL(sql1);
%>
<%
// 取随机产生的认证码(4 位数字)
String sRand="";
//生成随机类
Random random = new Random();
```

```
for (int i=0;i<4;i++){
    String rand=String.valueOf(random.nextInt(10));
    sRand+=rand;
}
// 将认证码存入 SESSION
session.setAttribute("rand",sRand);
%>
<html>
    <head>
        <title>欢迎访问冰鱼论坛『冰鱼论坛管理系统』</title>
        <meta http-equiv="Content-Type" content="text/html; charset=gb2312">
        <script Language="Javascript" src="include/form.js"></Script>
    </head>
    <body topmargin="0" style="text-align: center">
        <table width="100%" height="60" border="1" bordercolorlight="#7777ff"
             bordercolordark="#7777ff" style="border-collapse:collapse">
            <tr bgcolor="#dff2ed">
                <td valign=middle align=center>
                <%if(ad2_img==null||ad2_img.equals("")){%>
                此处可放置广告(首页大幅广告)
                <%}else{%>
                <a href="<%=ad2_url%>">
                <img border="0" onload="javascript:if(this.width>979){this.width=979;}"
                    src="<%=ad2_img%>" title="<%=ad2_title%>">
                </a>
                <%}%>
                </td>
            </tr>
        </table>
<p>
        <table width="100%" border="1" bordercolorlight="#7777ff" bordercolordark="#7777ff"
            style="border-collapse:collapse">
            <tr>
                <td height="25" colspan="2" background="images/skin/1/bg_td.gif">
                </td>
            </tr>
            <tr bgcolor="#dff2ed" valign="middle" align="center">
                <td width="50%">
                    <%
                    request.setCharacterEncoding("gb2312");
                    if(user_login){
                    %>
                    <table height="100%" valign="middle" align="center" border="0">
                        <tr>
                            <td width="30%">
```

```
                <!--img src="<%=userBean.getUser_Face()%>"
                onload="javascript:if(this.width>=120){width=120;}"-->
                欢迎您登录!
            </td>
            <td>
            </td>
        </tr>
    </table>
    <%}else{%>
    <form method="post" name="userlogin" action="check.jsp"><br>
    <table height="100%" valign="middle" align="center" border="0">
        <tr>
            <td>欢迎访问 <b>冰鱼论坛</b> ,您还没有【
                <a class=yh href="reg.jsp">注册</a>】或【
                <a class=yh href="login.jsp">登录</a>】
            </td>
        </tr>
        <tr>
            <td>用户名:
                    <input type="text" name="user_name" size="13">
  
                验证码:
                <input name="checknumber" type="text" maxlength="4"
size="6">
                <%out.print(sRand);%>
            </td>
        </tr>
        <tr>
            <td>密 码:
                <input type="password" name="user_password" size="13">
 
                Cookie:
                <select name="save_login">
                    <option value="no_time" selected>不保存
                    <option value="one_day">保存一天
                    <option value="one_month">保存一月
                    <option value="one_year">保存一年
                </select>
            </td>
        </tr>
        <tr>
            <td>
            <input type="submit" onclick="CheckLogin()" value="登 录">
 
            <input type="reset" value="重 填">  
```

```
                        『 <a class=yh href="">忘记密码</a> 』
                        </td>
                    </tr>
                </table>
                </form>
                <%}%>
            </td>
            <td width="50%">论坛基本情况模块</td>
        </tr>
        <tr>
            <td bgcolor="#dff2ed" colspan="2" height="30">
            </td>
        </tr>
    </table> <p>
    <%while(rs1.next()){
    String board_id=rs1.getString("board_id");
    %>
    <table width="100%" border="1" bordercolorlight="#7777FF" bordercolordark="#7777FF"
        style="border-collapse:collapse">
        <tr>
            <td height="25" colspan="2" background="images/skin/1/bg_td.gif" valign=
"middle">
            <b>
    <font color="#ffffff"> 
    <img src="images/skin/1/nofollow.gif">
    <a class=xh href="board.jsp?bm=<%=rs1.getString("board_name")%>&
        bmid=<%=rs1.getString("board_id")%>&
        board_id=<%=rs1.getString("board_id")%>">
        <%=rs1.getString("board_name")%>
    </a>
    </font>
    </td>
</tr>
    <%
    String sql2="select * from icefish_board WHERE board_bid="+board_id+" order by
board_id asc";ResultSet rs2=null;
   String _board_todaynum="";
   String _board_topicnum="";
   String _board_postnum="";
        rs2=indexBean.executeSQL(sql2);
        while(rs2.next()){
        //
        _board_todaynum=rs2.getString("board_todaynum");
        if (_board_todaynum==null){
        _board_todaynum="0";
```

```
        }
        //
        _board_topicnum=rs2.getString("board_topicnum");
        if (_board_topicnum==null){
        _board_topicnum="0";
        }
        //
        _board_postnum=rs2.getString("board_postnum");
        if (_board_postnum==null){
        _board_postnum="0";
        }
        %>
    <tr>
        <td height="85" valign="middle" align="center" width="43">
        <img src="images/skin/1/forum_nonews.gif">
        </td>
        <td>
                <table width="100%" height="100%" border="2" bordercolorlight="#ffffff"
                    bordercolordark="#ffffff" style="border-collapse:collapse">
                    <tr>
                        <td valign="top">
                        <a class=zh href="board.jsp?bm=<%=rs1.getString("board_
name")%>&
                        bmid=<%=rs1.getString("board_id")%>&
                        board_id=<%=rs2.getString("board_id")%>">
                        <%=rs2.getString("board_name")%>
                        </a>
                        <br>
                        <img src="images/skin/1/forum_readme.gif">
                        <%=rs2.getString("board_info")%>
                        </td>
                        <td width="27%">主题：<br>发帖：<br>日期：</td>
                    </tr>
                        <tr bgcolor="#dff2ed" height="23">
                            <td>版主：
                            <%=rs2.getString("board_master")%>
                            </td>
                            <td>
                            <font color="#ff0000">
                            <img src="images/skin/1/forum_today.gif">
                            <%=_board_todaynum%>
                            </font><font color="#0000ff">
                            <img src="images/skin/1/forum_topic.gif">
```

```
                        <%=_board_topicnum%>
                        <img src="images/skin/1/forum_post.gif">
                        <%=_board_postnum%></font></td>
                    </tr>
                </table>
            </td>
        </tr>
    <%}
    rs2.close();
    %>
</table> <p>
<%}
rs1.close();
%>
<table width="100%" height="100" border="1" bordercolorlight="#7777ff"
    bordercolordark="#7777ff" style="border-collapse:collapse">
    <tr>
        <td height="25" colspan="2" background="images/skin/1/bg_td.gif">
        <b>
        <font color="#ffffff">
        => 友情论坛;
        <a class=xh href="linkadd.jsp">[申请友情链接]</a>
        </font>
        </td>
    </tr>
    <tr bgcolor="#dff2ed"><td valign=middle align=center>
        <table width="100%" border="0">
            <tr align=left><td>
            <%
            String sql3="select * from icefish_link where link_islogo='false'
                    and link_pass='true'";
            ResultSet rs3=indexBean.executeSQL(sql3);
            while(rs3.next()){
            %>
            <a class=zh href="<%=rs3.getString("link_url")%>">
            <%=rs3.getString("link_name")%></a>
            <%}
            rs3.close();
            %>
            </td></tr>
            <tr><td><hr></hr></td></tr>
            <tr align=left><td>
            <%
            String sql4="select * from icefish_link where link_islogo='true' and link_
pass='true'";
```

```
                    ResultSet rs4=indexBean.executeSQL(sql4);
                    while(rs4.next()){
                    %>
                    <a class=zh href="<%=rs4.getString("link_url")%>">
                    <img border="0" width="88" height="31" src="<%=rs4.getString
("link_logo")%>"
                        title="<%=rs4.getString("link_name")%>">
                    </a>
                    <%}
                    rs3.close();
                    indexBean.close();
                    %>
                    </td></tr>
                </table>
            </td></tr>
        </table>
    </body>
</html>
```

12.3.3　组件与 Servlet 管理

BBS 论坛系统中使用的组件与 Servlet 包层次结构如图 12-3 所示。

1. bean 包

bean 包里存放的 Java 程序是实现论坛系统主要功能所要用到的数据模型和业务模型，包括用户数据、帖子数据、版块数据、友情链接数据以及数据库连接等方面。

2. edit 包

edit 包里存放的是实现论坛系统“在线编辑文本器”功能相关的数据模型和业务模型，包括设置样式、设置全屏幕选项、设置内容选项等。

3. page 包

page 包里存放的是结果集处理相关的业务逻辑，包括显示总页数、总记录行数、当前页号、当前页的记录条数、分页大小等。

4. servlet 包

servlet 包里存放的是论坛系统的控制器。具体控制器功能将在 12.5 节加以说明。

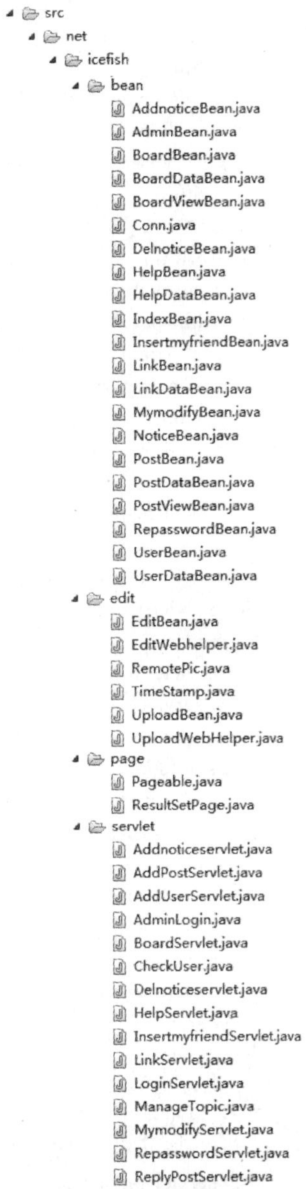

```
src
  net
    icefish
      bean
        AddnoticeBean.java
        AdminBean.java
        BoardBean.java
        BoardDataBean.java
        BoardViewBean.java
        Conn.java
        DelnoticeBean.java
        HelpBean.java
        HelpDataBean.java
        IndexBean.java
        InsertmyfriendBean.java
        LinkBean.java
        LinkDataBean.java
        MymodifyBean.java
        NoticeBean.java
        PostBean.java
        PostDataBean.java
        PostViewBean.java
        RepasswordBean.java
        UserBean.java
        UserDataBean.java
      edit
        EditBean.java
        EditWebhelper.java
        RemotePic.java
        TimeStamp.java
        UploadBean.java
        UploadWebHelper.java
      page
        Pageable.java
        ResultSetPage.java
      servlet
        Addnoticeservlet.java
        AddPostServlet.java
        AddUserServlet.java
        AdminLogin.java
        BoardServlet.java
        CheckUser.java
        Delnoticeservlet.java
        HelpServlet.java
        InsertmyfriendServlet.java
        LinkServlet.java
        LoginServlet.java
        ManageTopic.java
        MymodifyServlet.java
        RepasswordServlet.java
        ReplyPostServlet.java
```

图 12-3　包层次结构图

12.3.4 配置文件管理

在 icefish 的配置文件 web. xml 中，部署了 BBS 论坛系统用到的控制器。

代码模板 web. xml 如下：

```
<?xml version="1.0" encoding="UTF-8"?>
<!DOCTYPE web-app PUBLIC "-//Sun Microsystems, Inc.//DTD Web Application 2.3//EN"
"http://java.sun.com/dtd/web-app_2_3.dtd">
<web-app>
  <display-name>defaultroot</display-name>
  <servlet>
    <servlet-name>debugjsp</servlet-name>
<description>
Added to compile JSPs with debug info
</description>
    <servlet-class>org.apache.jasper.servlet.JspServlet</servlet-class>
    <init-param>
      <param-name>classdebuginfo</param-name>
      <param-value>true</param-value>
    </init-param>
    <load-on-startup>3</load-on-startup>
  </servlet>
  <servlet-mapping>
    <servlet-name>debugjsp</servlet-name>
    <url-pattern> *.jsp</url-pattern>
  </servlet-mapping>

  <servlet>
    <servlet-name>AddPostServlet</servlet-name>
    <servlet-class>net.icefish.servlet.AddPostServlet</servlet-class>
  </servlet>
  <servlet-mapping>
    <servlet-name>AddPostServlet</servlet-name>
    <url-pattern>/AddPostServlet</url-pattern>
  </servlet-mapping>

  <servlet>
    <servlet-name>EditPostServlet</servlet-name>
    <servlet-class>net.icefish.servlet.EditPostServlet</servlet-class>
  </servlet>
  <servlet-mapping>
    <servlet-name>EditPostServlet</servlet-name>
    <url-pattern>/EditPostServlet</url-pattern>
  </servlet-mapping>
```

```
<servlet>
  <servlet-name>ReplyPostServlet</servlet-name>
  <servlet-class>net.icefish.servlet.ReplyPostServlet</servlet-class>
</servlet>
<servlet-mapping>
  <servlet-name>ReplyPostServlet</servlet-name>
  <url-pattern>/ReplyPostServlet</url-pattern>
</servlet-mapping>

<servlet>
  <servlet-name>AddUserServlet</servlet-name>
  <servlet-class>net.icefish.servlet.AddUserServlet</servlet-class>
</servlet>
<servlet-mapping>
  <servlet-name>AddUserServlet</servlet-name>
  <url-pattern>/AddUserServlet</url-pattern>
</servlet-mapping>

<servlet>
  <servlet-name>LoginServlet</servlet-name>
  <servlet-class>net.icefish.servlet.LoginServlet</servlet-class>
</servlet>
<servlet-mapping>
  <servlet-name>LoginServlet</servlet-name>
  <url-pattern>/LoginServlet</url-pattern>
</servlet-mapping>

<servlet>
  <servlet-name>CheckUser</servlet-name>
  <servlet-class>net.icefish.servlet.CheckUser</servlet-class>
</servlet>
<servlet-mapping>
  <servlet-name>CheckUser</servlet-name>
  <url-pattern>/CheckUser</url-pattern>
</servlet-mapping>

<servlet>
  <servlet-name>AdminLogin</servlet-name>
  <servlet-class>net.icefish.servlet.AdminLogin</servlet-class>
</servlet>
<servlet-mapping>
  <servlet-name>AdminLogin</servlet-name>
  <url-pattern>/admin/AdminLogin</url-pattern>
</servlet-mapping>
```

```
<servlet>
  <servlet-name>BoardServlet</servlet-name>
  <servlet-class>net.icefish.servlet.BoardServlet</servlet-class>
</servlet>
<servlet-mapping>
  <servlet-name>BoardServlet</servlet-name>
  <url-pattern>/admin/BoardServlet</url-pattern>
</servlet-mapping>

<servlet>
  <servlet-name>HelpServlet</servlet-name>
  <servlet-class>net.icefish.servlet.HelpServlet</servlet-class>
</servlet>
<servlet-mapping>
  <servlet-name>HelpServlet</servlet-name>
  <url-pattern>/admin/HelpServlet</url-pattern>
</servlet-mapping>

<servlet>
  <servlet-name>ADServlet</servlet-name>
  <servlet-class>net.icefish.servlet.ADServlet</servlet-class>
</servlet>
<servlet-mapping>
  <servlet-name>ADServlet</servlet-name>
  <url-pattern>/admin/ADServlet</url-pattern>
</servlet-mapping>

<servlet>
  <servlet-name>LinkServlet</servlet-name>
  <servlet-class>net.icefish.servlet.LinkServlet</servlet-class>
</servlet>
<servlet-mapping>
  <servlet-name>LinkServlet</servlet-name>
  <url-pattern>/admin/LinkServlet</url-pattern>
</servlet-mapping>

<servlet>
  <servlet-name>Addnoticeservlet</servlet-name>
  <servlet-class>net.icefish.servlet.Addnoticeservlet</servlet-class>
</servlet>
<servlet-mapping>
  <servlet-name>Addnoticeservlet</servlet-name>
  <url-pattern>/Addnoticeservlet</url-pattern>
</servlet-mapping>
```

```
<servlet>
  <servlet-name>Delnoticeservlet</servlet-name>
  <servlet-class>net.icefish.servlet.Delnoticeservlet</servlet-class>
</servlet>
<servlet-mapping>
  <servlet-name>Delnoticeservlet</servlet-name>
  <url-pattern>/Delnoticeservlet</url-pattern>
</servlet-mapping>

<servlet>
  <servlet-name>MymodifyServlet</servlet-name>
  <servlet-class>net.icefish.servlet.MymodifyServlet</servlet-class>
</servlet>
<servlet-mapping>
  <servlet-name>MymodifyServlet</servlet-name>
  <url-pattern>/MymodifyServlet</url-pattern>
</servlet-mapping>

<servlet>
  <servlet-name>InsertmyfriendServlet</servlet-name>
  <servlet-class>net.icefish.servlet.InsertmyfriendServlet</servlet-class>
</servlet>
<servlet-mapping>
  <servlet-name>InsertmyfriendServlet</servlet-name>
  <url-pattern>/InsertmyfriendServlet</url-pattern>
</servlet-mapping>

<servlet>
  <servlet-name>RepasswordServlet</servlet-name>
  <servlet-class>net.icefish.servlet.RepasswordServlet</servlet-class>
</servlet>
<servlet-mapping>
  <servlet-name>RepasswordServlet</servlet-name>
  <url-pattern>/RepasswordServlet</url-pattern>
</servlet-mapping>

<servlet>
  <servlet-name>ManageTopic</servlet-name>
  <servlet-class>net.icefish.servlet.ManageTopic</servlet-class>
</servlet>
<servlet-mapping>
  <servlet-name>ManageTopic</servlet-name>
  <url-pattern>/managetopic</url-pattern>
</servlet-mapping>

</web-app>
```

12.4 组件设计

本系统中用到的组件有数据库连接与关闭、实体模型(数据封装 bean)和业务模型(业务处理 bean)。

12.4.1 数据库连接与关闭

数据库的连接与关闭是由 bean 包中的 Conn 类实现的,类中主要有两个方法:public static Connection connection()方法负责连接,public void close()方法负责关闭。

代码模板 Conn.java 如下:

```
package net.icefish.bean;
import java.sql.*;
public class Conn{
        public static Connection connection(){
        Connection conn=null;
        try{
            Class.forName("com.mysql.jdbc.Driver").newInstance();
        }
        catch(Exception e){
            System.out.print(e.toString());
        }

        try{
              String url="jdbc:mysql://127.0.0.1:3306/icefish?useUnicode=true&characterEncoding=
gb2312";
        //数据库名统一为 icefish,表名与字段名参照系统分析与设计
            String user="root";                              //设置用户名
            String password="admin";                         //设置密码
            conn=DriverManager.getConnection(url,user,password);
        }
        catch(SQLException e){
            System.out.print(e.toString());
        }
        return conn;

    }

    public void close(){
        try{
            conn.close();
        }
        catch(SQLException e){
```

```
                System.out.print(e.toString());
            }
        }
    }
```

12.4.2　实体模型

实体模型主要用于封装JSP页面提交的信息以及与数据库交互的数据传递。本系统共使用了七个实体模型：管理员(AdminBean)、用户(UserBean)、版块(BoardBean)、帖子(PostBean)、公告(NoticeBean)、帮助(HelpBean)和友情链接(LinkBean)。

代码模板AdminBean.java如下：

```
package net.icefish.bean;

public class AdminBean {
 private String admin_name;
 private String admin_password;
 private String admin_user;
 private boolean admin_login;

 public AdminBean(){
     admin_name=null;
     admin_password=null;
     admin_user=null;
     admin_login=false;
 }

 public String getAdmin_Name(){
     return admin_name;
 }
 public void setAdmin_Name(String admin_name){
     this.admin_name=admin_name;
 }

 public String getAdmin_Password(){
     return admin_password;
 }
 public void setAdmin_Password(String admin_password){
     this.admin_password=admin_password;
 }

 public String getAdmin_User(){
     return admin_user;
 }
 public void setAdmin_User(String admin_user){
```

```
        this.admin_user=admin_user;
    }

    public boolean getAdmin_Login(){
        return admin_login;
    }
    public void setAdmin_Login(boolean admin_login){
        this.admin_login=admin_login;
    }
}
```

代码模板 UserBean.java 如下：

```
package net.icefish.bean;
public class UserBean {
    private int user_id;
    private String user_name;
    private String user_password;
    private String user_password_q;
    private String user_password_a;
    private String user_sex;
    private String user_age;
    private String user_QQ;
    private String user_birthday;
    private String user_Email;
    private String user_tel;
    private String user_face;
    private String user_sign;
    private String user_grade;
    private String user_mark;
    private String user_topic;
    private String user_wealth;
    private String user_group;
    private String user_post;
    private String user_lastip;
    private String user_delnum;
    private String user_friends;
    private String user_regtime;
    private String user_lasttime;
    private boolean user_admin;
    private boolean user_locked;
    private boolean user_login;

    public UserBean(){
        user_mark="0";
        user_topic="0";
```

```
        user_post="0";
        user_wealth="0";
        user_delnum="0";
        user_login=false;
    }
    public int getUser_ID(){
        return user_id;
    }
    public void setUser_ID(int user_id){
        this.user_id=user_id;
    }
    public String getUser_Name(){
        return user_name;
    }
    public void setUser_Name(String user_name){
        this.user_name=user_name;
    }
    public String getUser_Password(){
        return user_password;
    }
    public void setUser_Password(String user_password){
        this.user_password=user_password;
    }
    public String getUser_Password_a(){
        return user_password_a;
    }
    public void setUser_Password_a(String user_password_a){
        this.user_password_a=user_password_a;
    }
    public String getUser_Password_q(){
        return user_password_q;
    }
    public void setUser_Password_q(String user_password_q){
        this.user_password_q=user_password_q;
    }
    public String getUser_Age(){
        return user_age;
    }
    public void setUser_Age(String user_age){
        this.user_age=user_age;
    }
    public String getUser_Sex(){
        return user_sex;
    }
    public void setUser_Sex(String user_sex){
```

```
        this.user_sex=user_sex;
    }
    public String getUser_QQ(){
        return user_QQ;
    }
    public void setUser_QQ(String user_QQ){
        this.user_QQ=user_QQ;
    }
    public String getUser_Birthday(){
        return user_birthday;
    }
    public void setUser_Birthday(String user_birthday){
        this.user_birthday=user_birthday;
    }
    public String getUser_Email(){
        return user_Email;
    }
    public void setUser_Email(String user_Email){
        this.user_Email=user_Email;
    }
    public String getUser_Face(){
        return user_face;
    }
    public void setUser_Face(String user_face){
        this.user_face=user_face;
    }
    public String getUser_Tel(){
        return user_tel;
    }
    public void setUser_Tel(String user_tel){
        this.user_tel=user_tel;
    }
    public String getUser_Sign(){
        return user_sign;
    }
    public void setUser_Sign(String user_sign){
        this.user_sign=user_sign;
    }
    public String getUser_Grade(){
        return user_grade;
    }
    public void setUser_Grade(String user_grade){
        this.user_grade=user_grade;
    }
    public String getUser_Topic(){
```

```
        return user_topic;
    }
    public void setUser_Topic(String user_topic){
        this.user_topic=user_topic;
    }
    public String getUser_Wealth(){
        return user_wealth;
    }
    public void setUser_Wealth(String user_wealth){
        this.user_wealth=user_wealth;
    }
    public String getUser_Mark(){
        return user_mark;
    }
    public void setUser_Mark(String user_mark){
        this.user_mark=user_mark;
    }
    public String getUser_Group(){
        return user_group;
    }
    public void setUser_Group(String user_group){
        this.user_group=user_group;
    }
    public String getUser_Post(){
        return user_post;
    }
    public void setUser_Post(String user_post){
        this.user_post=user_post;
    }
    public String getUser_LastIP(){
        return user_lastip;
    }
    public void setUser_LastIP(String user_lastip){
        this.user_lastip=user_lastip;
    }
    public String getUser_Delnum(){
        return user_delnum;
    }
    public void setUser_Delnum(String user_delnum){
        this.user_delnum=user_delnum;
    }
    public String getUser_Friends(){
        return user_friends;
    }
    public void setUser_Friends(String user_friends){
```

```
        this.user_friends=user_friends;
    }
    public String getUser_Regtime(){
        return user_regtime;
    }
    public void setUser_Regtime(String user_regtime){
        this.user_regtime=user_regtime;
    }
    public String getUser_Lasttime(){
        return user_lasttime;
    }
    public void setUser_Lasttime(String user_lasttime){
        this.user_lasttime=user_lasttime;
    }
    public boolean getUser_Admin(){
        return user_admin;
    }
    public void setUser_Admin(boolean user_admin){
        this.user_admin=user_admin;
    }
    public boolean getUser_Locked(){
        return user_locked;
    }
    public void setUser_Locked(boolean user_locked){
        this.user_locked=user_locked;
    }
    public boolean getUser_Login(){
        return user_login;
    }
    public void setUser_Login(boolean user_login){
        this.user_login=user_login;
    }
}
```

代码模板 BoardBean.java 如下：

```
public class BoardBean {
  private String board_name;
  private String board_info;
  private String board_id;
  private String board_bid;
  private String board_master;
  private boolean board_isMother;
  private String board_postnum;
  private String board_topicnum;
  private String board_lastreply;
```

```
private String board_todaynum;
private String board_img;

public BoardBean(){
    board_name=null;
    board_info=null;
    board_master=null;
    board_bid=null;
    board_isMother=true;
}
public String getBoard_ID(){
    return board_id;
}
public void setBoard_ID(String board_id){
    this.board_id=board_id;
}
public String getBoard_Name(){
    return board_name;
}
public void setBoard_Name(String board_name){
    this.board_name=board_name;
}
public String getBoard_Info(){
    return board_info;
}
public void setBoard_Info(String board_info){
    this.board_info=board_info;
}
public String getBoard_BID(){
    return board_bid;
}
public void setBoard_BID(String board_bid){
    this.board_bid=board_bid;
}
public String getBoard_Master(){
    return board_master;
}
public void setBoard_Master(String board_master){
    this.board_master=board_master;
}
public boolean getBoard_IsMother(){
    return board_isMother;
}
public void setBoard_IsMother(boolean board_isMother){
    this.board_isMother=board_isMother;
```

```
    }
    public String getBoard_Postnum(){
        return board_postnum;
    }
    public void setBoard_Postnum(String board_postnum){
        this.board_postnum=board_postnum;
    }
    public String getBoard_Topicnum(){
        return board_topicnum;
    }
    public void setBoard_Topicnum(String board_topicnum){
        this.board_topicnum=board_topicnum;
    }
    public String getBoard_Todaynum(){
        return board_todaynum;
    }
    public void setBoard_Todaynum(String board_todaynum){
        this.board_todaynum=board_todaynum;
    }
    public String getBoard_Lastreply(){
        return board_lastreply;
    }
    public void setBoard_Lastreply(String board_lastreply){
        this.board_lastreply=board_lastreply;
    }
    public String getBoard_Img(){
        return board_img;
    }
    public void setBoard_Img(String board_img){
        this.board_img=board_img;
    }
}
```

代码模板 PostBean.java 如下：

```
package net.icefish.bean;
public class PostBean {
  private String board_id;
  private String topic_id;
  private String post_content;
  private String topic_name;
  private String user_name;

  public PostBean(){
  }
  public String getBoard_Id(){
```

```
        return board_id;
    }
    public void setBoard_Id(String board_id){
        this.board_id=board_id;
    }
    public String getTopic_Id(){
        return topic_id;
    }
    public void setTopic_Id(String topic_id){
        this.topic_id=topic_id;
    }
    public String getPost_Content(){
        return post_content;
    }
    public void setPost_Content(String post_content){
        this.post_content=post_content;
    }
    public String getTopic_Name(){
        return topic_name;
    }
    public void setTopic_Name(String topic_name){
        this.topic_name=topic_name;
    }
    public String getUser_Name(){
        return user_name;
    }
    public void setUser_Name(String user_name){
        this.user_name=user_name;
    }
}
```

代码模板 NoticeBean.java 如下：

```
public class NoticeBean {
private String notice_id;
private String notice_author;
private String board_name;
private String notice_music;
private String notice_title;
private String notice_date;
private String notice_content;
private String xg;
public NoticeBean()
{
}
public void setNotice_id(String notice_id)
```

```
{
    this.notice_id = notice_id;
}
public void setCheckbox(String xg)
{
    this.xg = xg;
}
public void setNotice_Author(String notice_author)
{
    this.notice_author = notice_author;
}
public void setBoard_Name(String board_name)
{
    this.board_name = board_name;
}
public void setNotice_Music(String notice_music)
{
    this.notice_music = notice_music;
}

public void setNotice_Title(String notice_title)
{
    this.notice_title = notice_title;
}
public void setNotice_Date(String notice_date)
{
    this.notice_date = notice_date;
}
public void setNotice_Content(String notice_content)
{
    this.notice_content = notice_content;
}
public String getNotice_ID()
{
    return notice_id;
}
public String getNotice_Author()
{
    return notice_author;
}
public String getBoard_Name()
{
    return board_name;
}
public String getNotice_Music()
```

```
{
    return notice_music;
}
public String getNotice_Title()
{
    return notice_title;
}
public String getNotice_Date()
{
    return notice_date;
}
public String getNotice_Content()
{
    return notice_content;
}
public String getCheckbox()
{
    return xg;
}
}
```

代码模板 HelpBean.java 如下：

```
public class HelpBean {
 private String help_id;
 private String help_name;
 private String help_info;

 public HelpBean(){
     help_name=null;
     help_info=null;
 }
 public String getHelp_ID(){
     return help_id;
 }
 public void setHelp_ID(String help_id){
     this.help_id=help_id;
 }
 public String getHelp_Name(){
     return help_name;
 }
 public void setHelp_Name(String help_name){
     this.help_name=help_name;
 }
 public String getHelp_Info(){
     return help_info;
```

```
    }
    public void setHelp_Info(String help_info){
        this.help_info=help_info;
    }
}
```

代码模板 LinkBean.java 如下：

```
public class LinkBean {
    private String link_name;
    private String link_url;
    private String link_id;
    private boolean link_islogo;
    private boolean link_pass;
    private String link_info;
    private String link_logo;

    public LinkBean(){
        link_pass=false;
        link_islogo=false;
    }
    public String getLink_ID(){
        return link_id;
    }
    public void setLink_ID(String link_id){
        this.link_id=link_id;
    }
    public String getLink_Name(){
        return link_name;
    }
    public void setLink_Name(String link_name){
        this.link_name=link_name;
    }
    public String getLink_Info(){
        return link_info;
    }
    public void setLink_Info(String link_info){
        this.link_info=link_info;
    }
    public String getLink_Logo(){
        return link_logo;
    }
    public void setLink_Logo(String link_logo){
        this.link_logo=link_logo;
    }
    public String getLink_Url(){
```

```
        return link_url;
    }
    public void setLink_Url(String link_url){
        this.link_url=link_url;
    }
    public boolean getLink_Pass(){
        return link_pass;
    }
    public void setLink_Pass(boolean link_pass){
        this.link_pass=link_pass;
    }
    public boolean getLink_Islogo(){
        return link_islogo;
    }
    public void setLink_Islogo(boolean link_islogo){
        this.link_islogo=link_islogo;
    }
}
```

12.4.3 业务模型

本系统中主要用到的业务模型有如下几个：UserDataBean、BoardDataBean、BoardViewBean、PostDataBean、PostViewBean、AddnoticeBean、DelnoticeBean、HelpDataBean、InsertmyfriendBean、MymodifyBean、RepasswordBean、LinkDataBean。

UserDataBean业务模型的主要功能是：新用户注册并在注册时验证用户是否已存在；用户登录并在登录时验证用户名与密码是否能通过。

代码模板UserDataBean.java如下：

```
public class UserDataBean {
 private Connection conn;
 public UserDataBean(){
      this.conn=Conn.connection();
 }
 //注册时验证用户是否已存在
 public boolean checkUser(UserBean userBean){
      boolean flag=false;
      String user_name=userBean.getUser_Name();
      Statement stmt=null;
      try{
           stmt=conn.createStatement();
           ResultSet rs=stmt.executeQuery("select * from icefish_user where
                user_name='"+user_name+"'");
           if(!rs.next()){
           flag=true;
```

```
            }else{
                flag=false;
            }
            rs.close();
            stmt.close();
            conn.close();
        }
        catch(SQLException e){
            flag=false;
            System.out.println(e.toString());
        }
        return flag;
    }
    //登录时验证用户名与密码是否能通过
    public boolean loginUser(UserBean userBean){
        boolean flag=false;
        String user_name=userBean.getUser_Name();
        String user_password=userBean.getUser_Password();
        Statement stmt=null;
        try{
            stmt=conn.createStatement();
            ResultSet rs=stmt.executeQuery("select * from icefish_user where user_name='"+user_
name+"' and user_password='"+user_password+"'");
            if(!rs.next()){
            flag=false;
            }else{
                while(rs.next()){
                    String QQ=rs.getString("user_QQ");
                    userBean.setUser_Admin(rs.getBoolean("user_admin"));
                    userBean.setUser_Age(rs.getString("user_age"));
                    userBean.setUser_Birthday(rs.getString("user_birthday"));
                    userBean.setUser_Delnum(rs.getString("user_delnum"));
                    userBean.setUser_Email(rs.getString("user_Email"));
                    userBean.setUser_Face(rs.getString("user_face"));
                    userBean.setUser_Friends(rs.getString("user_friends"));
                    userBean.setUser_Grade(rs.getString("user_grade"));
                    userBean.setUser_Group(rs.getString("user_group"));
                    userBean.setUser_ID(rs.getInt("user_id"));
                    userBean.setUser_LastIP(rs.getString("user_lastip"));
                    userBean.setUser_Lasttime(rs.getString("user_lasttime"));
                    userBean.setUser_Locked(rs.getBoolean("user_locked"));
                    userBean.setUser_Mark(rs.getString("user_mark"));
                    userBean.setUser_Password_a(rs.getString("user_password_a"));
                    userBean.setUser_Password_q(rs.getString("user_password_q"));
                    userBean.setUser_Post(rs.getString("user_post"));
```

```
                userBean.setUser_QQ(QQ);
                userBean.setUser_Regtime(rs.getString("user_regtime"));
                userBean.setUser_Sex(rs.getString("user_sex"));
                userBean.setUser_Sign(rs.getString("user_sign"));
                userBean.setUser_Tel(rs.getString("user_tel"));
                userBean.setUser_Topic(rs.getString("user_topic"));
                userBean.setUser_Wealth(rs.getString("user_wealth"));

            }
            flag=true;
        }
        rs.close();
        stmt.close();
        conn.close();
    }
    catch(SQLException e){
        flag=false;
        System.out.println(e.toString());
    }

    return flag;

}
//后台管理登录验证
public boolean adminLogin(AdminBean adminBean){
    boolean flag=false;
    String admin_name=adminBean.getAdmin_Name();
    String admin_password=adminBean.getAdmin_Password();
    String admin_user=adminBean.getAdmin_User();
    Statement stmt=null;
    try{
        stmt=conn.createStatement();
        ResultSet rs=stmt.executeQuery("select * from icefish_admin where admin_name='"+
admin_name+"' and admin_password='"+admin_password+"' and admin_user='"+admin_user+"'");
        if(!rs.next()){
        flag=false;
        }else{
            flag=true;
        }
        rs.close();
        stmt.close();
        conn.close();
    }
    catch(SQLException e){
        flag=false;
```

```
            System.out.println(e.toString());
        }

        return flag;
    }
    //新用户注册
    public boolean addUser(UserBean userBean){
        boolean flag=false;

        PreparedStatement pstmt1=null;
        try{
            pstmt1=conn.prepareStatement("insert into icefish_user(
user_name,user_password,user_password_q,user_password_a,user_sex,user_Email,user_mark,user_topic,
user_wealth,user_post,user_delnum,user_regtime) values(?,?,?,?,?,?,?,?,?,?,?,now())");
            pstmt1.setString(1, userBean.getUser_Name());
            pstmt1.setString(2, userBean.getUser_Password());
            pstmt1.setString(3, userBean.getUser_Password_q());
            pstmt1.setString(4, userBean.getUser_Password_a());
            pstmt1.setString(5, userBean.getUser_Sex());
            pstmt1.setString(6, userBean.getUser_Email());
            pstmt1.setInt(7, 0);
            pstmt1.setInt(8, 0);
            pstmt1.setInt(9, 0);
            pstmt1.setInt(10, 0);
            pstmt1.setInt(11, 0);
            int result1=pstmt1.executeUpdate();
            if (result1 > 0){
                flag = true;
            }
            else{
                flag = false;
            }
            pstmt1.close();
            conn.close();
        }
        catch(SQLException e){
            flag=false;
            System.out.println(e.toString());
        }
        return flag;
    }
}
```

BoardDataBean 业务模型的主要功能是：增加版块、修改版块和删除版块。

代码模板 BoardDataBean.java 如下：

```
public class BoardDataBean {
 private Connection conn;

 public BoardDataBean(){
      this.conn=Conn.connection();
 }
 //增加新版块
 public boolean addBoard(BoardBean boardBean){
      boolean flag=false;

      boolean board_isMother=boardBean.getBoard_IsMother();
      if(board_isMother){
           PreparedStatement pstmt1=null;
           try{
                pstmt1=conn.prepareStatement("insert into icefish_board
(board_name, board_info, board_isMother, board_master, board_postnum, board_topicnum, board_
todaynum) values(?,?,'true',?,0,0,0)");
                pstmt1.setString(1, boardBean.getBoard_Name());
                pstmt1.setString(2, boardBean.getBoard_Info());
                pstmt1.setString(3, boardBean.getBoard_Master());
                int result1=pstmt1.executeUpdate();
                if (result1 > 0){
                     flag = true;
                }
                else{
                     flag = false;
                }
                pstmt1.close();
                conn.close();
           }
           catch(SQLException e){
                flag=false;
                System.out.println(e.toString());
           }
      }else{
           PreparedStatement pstmt1=null;
           try{
                pstmt1=conn.prepareStatement("insert into icefish_board
(board_name,board_info,board_isMother,board_master,board_bid,board_postnum,board_topicnum,
board_todaynum) values(?,?,'false',?,?,0,0,0)");
                pstmt1.setString(1, boardBean.getBoard_Name());
                pstmt1.setString(2, boardBean.getBoard_Info());
                pstmt1.setString(3, boardBean.getBoard_Master());
                pstmt1.setString(4, boardBean.getBoard_BID());
                int result1=pstmt1.executeUpdate();
```

```
                if (result1 > 0){
                    flag = true;
                }
                else{
                    flag = false;
                }
                pstmt1.close();
                conn.close();
            }
            catch(SQLException e){
                flag=false;
                System.out.println(e.toString());
            }
        }

        return flag;
    }
    //修改版块信息
    public boolean editBoard(BoardBean boardBean){
        boolean flag=false;
        boolean board_isMother=boardBean.getBoard_IsMother();
        if(board_isMother){
            PreparedStatement pstmt1=null;
            try{
                pstmt1=conn.prepareStatement("update icefish_board set board_name=?, board_
info=?, board_isMother='true', board_master=?, board_postnum=0, board_topicnum=0, board_
todaynum=0 where board_id="+boardBean.getBoard_ID());
                pstmt1.setString(1, boardBean.getBoard_Name());
                pstmt1.setString(2, boardBean.getBoard_Info());
                pstmt1.setString(3, boardBean.getBoard_Master());
                int result1=pstmt1.executeUpdate();
                if (result1 > 0){
                    flag = true;
                }
                else{
                    flag = false;
                }
                pstmt1.close();
                conn.close();
            }
            catch(SQLException e){
                flag=false;
                System.out.println(e.toString());
            }
        }else{
```

```
            PreparedStatement pstmt1=null;
            try{
                pstmt1=conn.prepareStatement("update icefish_board set board_name=?, board_info=?, board_isMother='false', board_master=?, board_bid=?, board_postnum=0, board_topicnum=0, board_todaynum=0 where board_id="+boardBean.getBoard_ID());
                pstmt1.setString(1, boardBean.getBoard_Name());
                pstmt1.setString(2, boardBean.getBoard_Info());
                pstmt1.setString(3, boardBean.getBoard_Master());
                pstmt1.setString(4, boardBean.getBoard_BID());
                int result1=pstmt1.executeUpdate();
                if (result1 > 0){
                    flag = true;
                }
                else{
                    flag = false;
                }
                pstmt1.close();
                conn.close();
            }
            catch(SQLException e){
                flag=false;
                System.out.println(e.toString());
            }
        }
        return flag;
    }
    //删除版块
    public boolean delBoard(BoardBean boardBean){
         boolean flag=false;
        boolean board_isMother=boardBean.getBoard_IsMother();
        if(board_isMother){
//如果所删除的版块为论坛一级分类版块,则连同其下属的子版块也删除,包括所有帖子
            PreparedStatement pstmt1=null;
            PreparedStatement pstmt2=null;
            PreparedStatement pstmt3=null;
            PreparedStatement pstmt4=null;
            try{
                pstmt4=conn.prepareStatement("select * from icefish_board where board_bid=?");
                pstmt4.setString(1, boardBean.getBoard_ID());
                ResultSet rs4=pstmt4.executeQuery();
                while(rs4.next()){
                    PreparedStatement pstmt5=null;
                    pstmt5=conn.prepareStatement("delete from icefish_topic where topic_boardid=?");
                    pstmt5.setString(1, rs4.getString("board_id"));
                    int result5=pstmt5.executeUpdate();
```

```
                PreparedStatement pstmt6=null;
                pstmt6=conn.prepareStatement("delete from icefish_post where post_boardid=?");
                pstmt6.setString(1, rs4.getString("board_id"));
                int result6=pstmt6.executeUpdate();
                if (result6>0 && result5>0){
                    flag = true;
                }
                else{
                    flag = false;
                }
                pstmt5.close();
                pstmt6.close();
            }
            pstmt4.close();
            pstmt1 = conn.prepareStatement("delete from icefish_board where board_id=? or
board_bid=?");
            pstmt1.setString(1, boardBean.getBoard_ID());
            pstmt1.setString(2, boardBean.getBoard_ID());
            int result1=pstmt1.executeUpdate();
            if (result1 > 0){
                flag = true;
            }
            else{
                flag = false;
            }
            pstmt1.close();
            pstmt2=conn.prepareStatement("delete from icefish_topic where topic_boardid=?");
            pstmt2.setString(1, boardBean.getBoard_ID());
            int result2=pstmt2.executeUpdate();
            pstmt3=conn.prepareStatement("delete from icefish_post where post_boardid=?");
            pstmt3.setString(1, boardBean.getBoard_ID());
            int result3=pstmt3.executeUpdate();
            if (result3>0 && result2>0){
                flag = true;
            }
            else{
                flag = false;
            }
            pstmt2.close();
            pstmt3.close();
            conn.close();
        }
        catch(SQLException e){
            flag=false;
            System.out.println(e.toString());
```

```
            }
        }else{
            PreparedStatement pstmt1=null;
            PreparedStatement pstmt2=null;
            PreparedStatement pstmt3=null;
            try{
                pstmt1=conn.prepareStatement("delete from icefish_board where board_id=?");
                pstmt1.setString(1, boardBean.getBoard_ID());
                int result1=pstmt1.executeUpdate();
                if (result1 > 0){
                    flag = true;
                }
                else{
                    flag = false;
                }
                pstmt1.close();
                pstmt2=conn.prepareStatement("delete from icefish_topic where topic_boardid=?");
                pstmt2.setString(1, boardBean.getBoard_ID());
                int result2=pstmt2.executeUpdate();
                if (result2 > 0){
                    flag = true;
                }
                else{
                    flag = false;
                }
                pstmt2.close();
                pstmt3=conn.prepareStatement("delete from icefish_post where post_boardid=?");
                pstmt3.setString(1, boardBean.getBoard_ID());
                int result3=pstmt3.executeUpdate();
                if (result3 > 0){
                    flag = true;
                }
                else{
                    flag = false;
                }
                pstmt3.close();
                conn.close();
            }
            catch(SQLException e){
                flag=false;
                System.out.println(e.toString());
            }
        }
        return flag;
    }
}
```

BoardViewBean 业务模型的主要功能是：浏览版块。

代码模板 BoardViewBean.java 如下：

```
public class BoardViewBean {
 private Connection conn;
 private ResultSet rs;

 public BoardViewBean(){
     this.conn=Conn.connection();
 }
 public ResultSet executeSQL(String sql) {
     try{
         Statement stmt = conn.createStatement();              //语句接口
         rs = stmt.executeQuery(sql);                          //获得结果集
     }
     catch(SQLException e){
         System.out.print(e.toString());
     }
     //conn.close();
     return rs;
 }
 public void close(){
     try{
         conn.close();
     }
     catch(SQLException e){
         System.out.print(e.toString());
     }
 }
}
```

PostDataBean 业务模型的主要功能是：将相关信息存入帖子数据表和话题数据表。

代码模板 PostDataBean 如下：

```
public class PostDataBean {
 private Connection conn;

 public PostDataBean(){
     this.conn=Conn.connection();
 }
 public boolean addPost(PostBean postBean){
     boolean flag=false;
     Statement stmt=null;
     PreparedStatement pstmt1=null;
     PreparedStatement pstmt2=null;
```

```
    try{
        //把相关信息存入表 icefish_topic,并获得当前插入的 topic 的 ID
        stmt=conn.createStatement();
        pstmt1=conn.prepareStatement("insert into icefish_topic
                (topic_boardid,topic_name,topic_user,topic_time) values(?,?,?,now())");
        pstmt1.setString(1, postBean.getBoard_Id());
        pstmt1.setString(2, postBean.getTopic_Name());
        pstmt1.setString(3, postBean.getUser_Name());
        int result1=pstmt1.executeUpdate();
        if (result1 > 0){
            flag = true;
        }
        else{
            flag = false;
        }
        ResultSet rs=stmt.executeQuery("select @@identity as 'topic_id'");
        while(rs.next()){
            String topic_id=rs.getString("topic_id");
            //把相关信息存入表 icefish_post
            pstmt2=conn.prepareStatement("insert into icefish_post
            (post_topicid,post_content,post_boardid,post_user,post_time) values(?,?,?,?,now()) ");
            pstmt2.setString(1, topic_id);
            pstmt2.setString(2, postBean.getPost_Content());
            pstmt2.setString(3, postBean.getBoard_Id());
            pstmt2.setString(4, postBean.getUser_Name());
            int result2=pstmt2.executeUpdate();
            if (result2 > 0){
                flag = true;
            }
            else{
                flag = false;
            }
            pstmt2.close();
        }
        rs.close();
        stmt.close();
        pstmt1.close();
        conn.close();
    }
    catch(SQLException e){
        flag=false;
        System.out.println(e.toString());
    }
    return flag;
}
```

```
    public boolean replyPost(PostBean postBean){
        boolean flag=false;
        PreparedStatement pstmt1=null;
        try{
                String topic_id=postBean.getTopic_Id();
                //把相关信息存入表 icefish_post
                pstmt1=conn.prepareStatement("insert into icefish_post
(post_topicid,post_content,post_boardid,post_user,post_time) values(?,?,?,?,now()) ");
                pstmt1.setString(1, topic_id);
                pstmt1.setString(2, postBean.getPost_Content());
                pstmt1.setString(3, postBean.getBoard_Id());
                pstmt1.setString(4, postBean.getUser_Name());
                int result2=pstmt1.executeUpdate();
                if (result2 > 0){
                    flag = true;
                }
                else{
                    flag = false;
                }
                pstmt1.close();
                pstmt1.close();
            conn.close();
        }
        catch(SQLException e){
            flag=false;
            System.out.println(e.toString());
        }
        return flag;
    }
}
```

PostViewBean 业务模型的主要功能是：浏览帖子。

代码模板 PostViewBean.java 如下：

```
public class PostViewBean {
 private Connection conn;
 private ResultSet rs;

 public PostViewBean(){
     this.conn=Conn.connection();
 }
 public ResultSet executeSQL(String sql) {
     try{
         Statement stmt = conn.createStatement();          //语句接口
         rs = stmt.executeQuery(sql);                      //获得结果集
     }
```

```
        catch(SQLException e){
            System.out.print(e.toString());
        }
        return rs;
    }
    public void close(){
        try{
            conn.close();
        }
        catch(SQLException e){
            System.out.print(e.toString());
        }
    }
}
```

AddnoticeBean 业务模型的主要功能是：增加公告。

代码模板 AddnoticeBean.java 如下：

```
public class AddnoticeBean {
 private Connection conn;

 public AddnoticeBean(){
      this.conn=Conn.connection();
 }
 public void ggBe(NoticeBean a){
      PreparedStatement pstmt=null;
      try{
      pstmt=conn.prepareStatement("insert into icefish_notice
(notice_ author, board _ name, notice _ music, notice _ title, notice _ content, notice _ date) values
(?,?,?,?,?,now())");
      pstmt.setString(1, a.getNotice_Author());
      pstmt.setString(2, a.getBoard_Name());
      pstmt.setString(3, a.getNotice_Music());
      pstmt.setString(4, a.getNotice_Title());
      pstmt.setString(5, a.getNotice_Content());
      pstmt.executeUpdate();
      pstmt.close();
      conn.close();
      }
      catch(SQLException e){
      System.out.println(e);
      }
 }
}
```

DelnoticeBean 业务模型的主要功能是：删除公告。

代码模板 DelnoticeBean.java 如下：

```
package net.icefish.bean;
import java.sql.Connection;
import java.sql.PreparedStatement;
import java.sql.SQLException;
import net.icefish.bean.Conn;
public class DelnoticeBean {
 private Connection conn;

 public DelnoticeBean(){
      this.conn=Conn.connection();
 }
 public void delBe(NoticeBean a){
      PreparedStatement pstmt=null;
      try{
      String xg=a.getNotice_ID();
      String sql="delete from icefish_notice where notice_id in("+xg+")";
                  pstmt.executeUpdate(sql);
                  pstmt.close();
                  conn.close();
              }
      catch(SQLException e){
      System.out.println(e);
      }
 }
}
```

HelpDataBean 业务模型的主要功能是：增加帮助、修改帮助和删除帮助。

代码模板 HelpDataBean.java 如下：

```
public class HelpDataBean {
 private Connection conn;

 public HelpDataBean(){
      this.conn=Conn.connection();
 }
 //增加新帮助
 public boolean addHelp(HelpBean helpBean){
      boolean flag=false;
      PreparedStatement pstmt1=null;
      try{
            pstmt1 = conn.prepareStatement("insert into icefish_help(help_name,help_info)
values(?,?)");
          pstmt1.setString(1, helpBean.getHelp_Name());
          pstmt1.setString(2, helpBean.getHelp_Info());
```

```
            int result1=pstmt1.executeUpdate();
            if (result1 > 0){
                flag = true;
            }
            else{
                flag = false;
            }
            pstmt1.close();
            conn.close();
        }
        catch(SQLException e){
            flag=false;
            System.out.println(e.toString());
        }
        return flag;
    }
    //修改帮助内容
    public boolean editHelp(HelpBean helpBean){
        boolean flag=false;
        PreparedStatement pstmt1=null;
        try{
            pstmt1 = conn.prepareStatement("update icefish_help set help_name=?, help_info=?
where help_id="+helpBean.getHelp_ID());
            pstmt1.setString(1, helpBean.getHelp_Name());
            pstmt1.setString(2, helpBean.getHelp_Info());
            int result1=pstmt1.executeUpdate();
            if (result1 > 0){
                flag = true;
            }
            else{
                flag = false;
            }
            pstmt1.close();
            conn.close();
        }
        catch(SQLException e){
            flag=false;
            System.out.println(e.toString());
        }
        return flag;
    }
    //删除帮助内容
    public boolean delHelp(HelpBean helpBean){
        boolean flag=false;
        PreparedStatement pstmt1=null;
```

```
        try{
            pstmt1=conn.prepareStatement("delete from icefish_help where help_id=?");
            pstmt1.setString(1, helpBean.getHelp_ID());
            int result1=pstmt1.executeUpdate();
            if (result1 > 0){
                flag = true;
            }
            else{
                flag = false;
            }
            pstmt1.close();
            conn.close();
        }
        catch(SQLException e){
            flag=false;
            System.out.println(e.toString());
        }
        return flag;
    }
}
```

InsertmyfriendBean 业务模型的主要功能是：在论坛中添加好友。

代码模板 InsertmyfriendBean.java 如下：

```
package net.icefish.bean;
import java.sql.*;
import net.icefish.bean.Conn;
public class InsertmyfriendBean {
 private Connection conn;

 public InsertmyfriendBean(){
     this.conn=Conn.connection();
 }
 public void tjBe(UserBean a){
     PreparedStatement pstmt=null;
     PreparedStatement pstmt2=null;
     try{
     pstmt=conn.prepareStatement("insert into icefish_friend
         (user_id,friend_id,friend_name,friend_addtime,friend_Email) value(?,?,?,now(),?)");
     pstmt.setString(1, a.getUser_ID());
     pstmt.setString(2, a.getFriend_ID());
     pstmt.setString(3, a.getFriend_Name());
     pstmt.setString(4, a.getFriend_Email());
     pstmt.executeUpdate();
     pstmt.close();
     conn.close();
```

```
        }
        catch(SQLException e){
        System.out.println(e);
        }
  }
}
```

MymodifyBean业务模型的主要功能是：用户个人信息修改。

代码模板 MymodifyBean.java 如下：

```
package net.icefish.bean;
import java.sql.*;
import net.icefish.bean.Conn;
public class MymodifyBean {
private Connection conn;

  public MymodifyBean(){
        this.conn=Conn.connection();
  }
        public void xgBe(UserBean a){
        PreparedStatement pstmt=null;
        try{
        pstmt=conn.prepareStatement("update icefish_user set user_secondname=?,user_sex=?,
        user_birthday=?,user_QQ=?,user_group=?,user_face=?,user_sign=?,user_truename=?,
user_Email=?,user_tel=?,user_blood=?,user_shengxiao=?,user_nation=?,user_province=?,
user_city=? where user_id=?");
        pstmt.setString(1, a.getUser_Secondname());
        pstmt.setString(2, a.getUser_Sex());
        pstmt.setString(3, a.getUser_Birthday());
        pstmt.setString(4, a.getUser_QQ());
        pstmt.setString(5, a.getUser_Group());
        pstmt.setString(6, a.getUser_Face());
        pstmt.setString(7, a.getUser_Sign());
        pstmt.setString(8, a.getUser_Truename());
        pstmt.setString(9, a.getUser_Email());
        pstmt.setString(10, a.getUser_Tel());
        pstmt.setString(11, a.getUser_Blood());
        pstmt.setString(12, a.getUser_Shengxiao());
        pstmt.setString(13, a.getUser_Nation());
        pstmt.setString(14, a.getUser_Province());
        pstmt.setString(15, a.getUser_City());
        pstmt.setString(16, a.getUser_ID());
        pstmt.executeUpdate();
        pstmt.close();
        conn.close();
        }
```

```
        catch(SQLException e){
        System.out.println(e);
        }
  }
}
```

RepasswordBean 业务模型的主要功能是：修改密码。

代码模板 RepasswordBean.java 如下：

```
import java.sql.*;
import net.icefish.bean.Conn;
public class RepasswordBean {
  private Connection conn;

  public RepasswordBean(){
        this.conn=Conn.connection();
  }
  public void mmBe(UserBean a){
        PreparedStatement pstmt=null;
        try{
        pstmt=conn.prepareStatement("update icefish_user set user_password=? where user_id=?");
        pstmt.setString(1, a.getUser_Password());
        pstmt.setString(2, a.getUser_ID());
        pstmt.executeUpdate();
        pstmt.close();
        conn.close();
        }
        catch(SQLException e){
        System.out.println(e);
        }
  }
}
```

LinkDataBean 业务模型的主要功能是：添加友情链接、修改友情链接和删除友情链接。

代码模板 LinkDataBean.java 如下：

```
public class LinkDataBean {
  private Connection conn;
  public LinkDataBean(){
        this.conn=Conn.connection();
  }
  //增加新链接
  public boolean addLink(LinkBean linkBean){
        boolean flag=false;
        String link_pass=new String();
```

```
        if(linkBean.getLink_Pass()){
            link_pass="true";
        }else{
            link_pass="false";
        }
        String link_islogo=new String();
        if(linkBean.getLink_Islogo()){
            link_islogo="true";
        }else{
            link_islogo="false";
        }
        PreparedStatement pstmt1=null;
        try{
            pstmt1=conn.prepareStatement("insert into icefish_link
(link_name,link_info,link_url,link_logo,link_islogo,link_pass) values(?,?,?,?,?,?)");
            pstmt1.setString(1, linkBean.getLink_Name());
            pstmt1.setString(2, linkBean.getLink_Info());
            pstmt1.setString(3, linkBean.getLink_Url());
            pstmt1.setString(4, linkBean.getLink_Logo());
            pstmt1.setString(5, link_islogo);
            pstmt1.setString(6, link_pass);
            int result1=pstmt1.executeUpdate();
            if (result1 > 0){
                flag = true;
            }
            else{
                flag = false;
            }
            pstmt1.close();
            conn.close();
        }
        catch(SQLException e){
            flag=false;
            System.out.println(e.toString());
        }
        return flag;
    }
    //修改链接内容
    public boolean editLink(LinkBean linkBean){
        boolean flag=false;
        String link_pass=new String();
        if(linkBean.getLink_Pass()){
            link_pass="true";
        }else{
            link_pass="false";
```

```
        }
        String link_islogo=new String();
        if(linkBean.getLink_Islogo()){
            link_islogo="true";
        }else{
            link_islogo="false";
        }
        PreparedStatement pstmt1=null;
        try{
            pstmt1=conn.prepareStatement("update icefish_link set link_name=?,link_info=?,
            link_url=?,link_logo=?,link_islogo=?,link_pass=? where link_id="+linkBean
    .getLink_ID());
            pstmt1.setString(1, linkBean.getLink_Name());
            pstmt1.setString(2, linkBean.getLink_Info());
            pstmt1.setString(3, linkBean.getLink_Url());
            pstmt1.setString(4, linkBean.getLink_Logo());
            pstmt1.setString(5, link_islogo);
            pstmt1.setString(6, link_pass);
            int result1=pstmt1.executeUpdate();
            if (result1 > 0){
                flag = true;
            }
            else{
                flag = false;
            }
            pstmt1.close();
            conn.close();
        }
        catch(SQLException e){
            flag=false;
            System.out.println(e.toString());
        }
        return flag;
    }
    //删除链接内容
    public boolean delLink(LinkBean linkBean){
        boolean flag=false;
        PreparedStatement pstmt1=null;
        try{
            pstmt1=conn.prepareStatement("delete from icefish_link where link_id=?");
            pstmt1.setString(1, linkBean.getLink_ID());
            int result1=pstmt1.executeUpdate();
            if (result1 > 0){
                flag = true;
            }
            else{
                flag = false;
            }
```

```
            pstmt1.close();
            conn.close();
        }
        catch(SQLException e){
            flag=false;
            System.out.println(e.toString());
        }
        return flag;
    }
}
```

12.5　系统实现

12.5.1　用户注册

当用户注册时,该模块要求用户接受注册声明,然后输入用户名、性别、密码等信息,否则不允许注册。注册成功的用户信息被存入数据库的 icefish_user 表中。

1. 视图(JSP 页面)

该模块中有三个 JSP 页面: reg.jsp、adduser.jsp 和 regsuccess.jsp。reg.jsp 显示注册说明,需要用户接受方可进行后续的注册; adduser.jsp 提供注册信息输入界面; regsucess.jsp 提示用户注册是否成功。

代码模板 reg.jsp 代码如下,效果如图 12-4 所示。

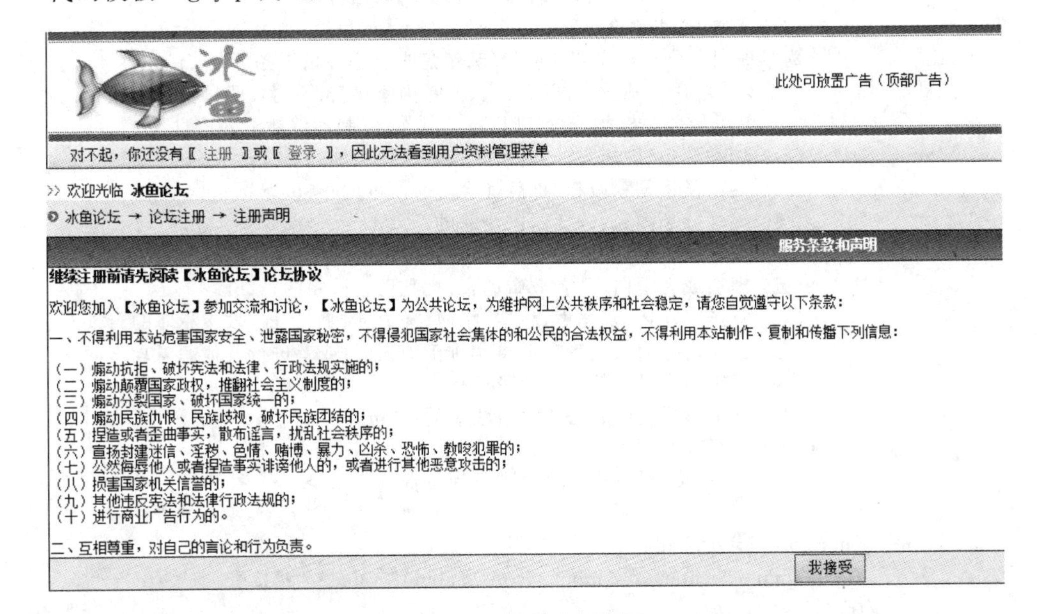

图 12-4　注册声明页面

```
<%@page pageEncoding="gb2312"%>
<%@page contentType="text/html;charset=gb2312" %>
<%@include file="top.jsp" %>
<html>
 <head>
     <title>冰鱼论坛----论坛注册条约阅读</title>
     <script Language="Javascript" src="include/form.js"></Script>
 </head>
 <body style="text-align: center">
     <div align=left>>> 欢迎光临 <b>冰鱼论坛</b></div>
      < table width = " 100%" height = " 30 " border = " 0 " bordercolorlight = " # 7777ff "
bordercolordark="#7777ff" style="border-collapse:collapse">
         <tr bgcolor="#dff2ed"><td><img src="images/skin/1/forum_nav.gif">
         <a class=zh href="">冰鱼论坛</a> →
             <a class=zh href="reg.jsp">论坛注册</a> → 注册声明</td></tr>
     </table>
     <form method="post" name="myform" action="adduser.jsp?reg=apply">
     < table width = "100%" border = "1" bordercolorlight = " # 7777ff" bordercolordark = " #
7777ff" style="border-collapse:collapse">
         <tr>
             <td height="25" align="center" background="images/skin/1/bg_td.gif"><b>
<font color="#ffffff">服务条款和声明</font></b></td>
         </tr>
         <tr>
             <td><b>继续注册前请先阅读【冰鱼论坛】论坛协议</b><p>
                 欢迎您加入【冰鱼论坛】参加交流和讨论,【冰鱼论坛】为公共论坛,
                 为维护网上公共秩序和社会稳定,请您自觉遵守以下条款:<p>
                 一、不得利用本站危害国家安全、泄露国家秘密,不得侵犯国家社会集体的和
                    公民的合法权益,不得利用本站制作、复制和传播下列信息:<p>
                 (一) 煽动抗拒、破坏宪法和法律、行政法规实施的;<br>
                 (二) 煽动颠覆国家政权,推翻社会主义制度的;<br>
                 (三) 煽动分裂国家、破坏国家统一的;<br>
                 (四) 煽动民族仇恨、民族歧视,破坏民族团结的;<br>
                 (五) 捏造或者歪曲事实,散布谣言,扰乱社会秩序的;<br>
                 (六) 宣扬封建迷信、淫秽、色情、赌博、暴力、凶杀、恐怖、教唆犯罪的;<br>
                 (七) 公然侮辱他人或者捏造事实诽谤他人的,或者进行其他恶意攻击的;<br>
                 (八) 损害国家机关信誉的;<br>
                 (九) 其他违反宪法和法律行政法规的;<br>
                 (十) 进行商业广告行为的. <p>
                 二、互相尊重,对自己的言论和行为负责. </td>
         </tr>
         <tr bgcolor="#dff2ed">
             <td align="center"><input type="submit" value="我接受"></td>
         </tr>
     </table>
```

```
    </form>
  </body>
</html>
```

代码模板 adduser.jsp 如下，效果如图 12-5 所示。

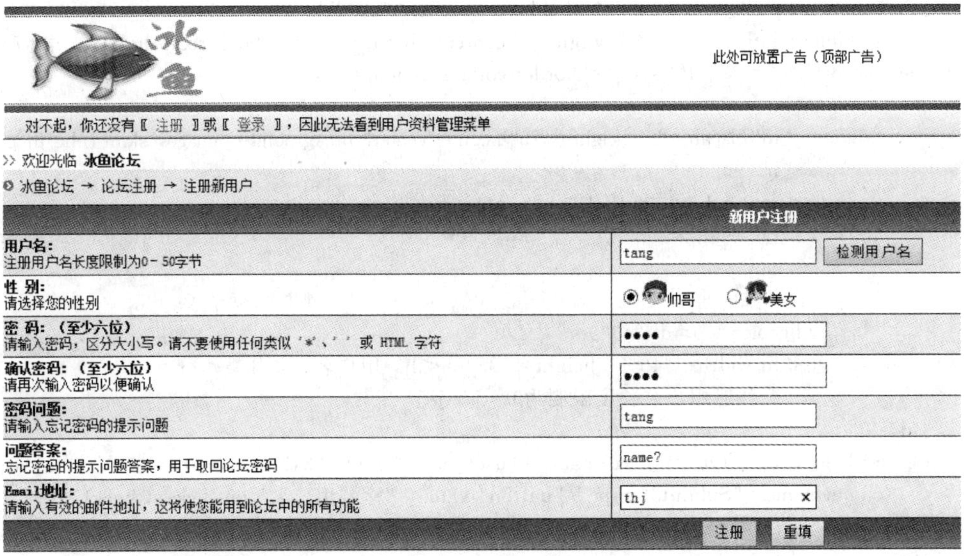

图 12-5　注册信息填写页面

```
<%@page pageEncoding="gb2312"%>
<%@page contentType="text/html;charset=gb2312" %>
<%@include file="top.jsp" %>
<html>
  <head>
      <title>冰鱼论坛----论坛注册新用户</title>
      <script Language="Javascript" src="include/form.js"></Script>
  </head>
  <body style="text-align: center">
      <div align=left>>> 欢迎光临 <b>冰鱼论坛</b></div>
       < table width = " 100%" height = " 30 " border = " 0 " bordercolorlight = " #7777ff"
bordercolordark="#7777ff" style="border-collapse:collapse">
          <tr bgcolor="#dff2ed">
          <td>
          <img src="images/skin/1/forum_nav.gif">
          <a class=zh href="">冰鱼论坛</a> →
          <a class=zh href="reg.jsp">论坛注册</a> → 注册新用户
          </td>
          </tr>
      </table>
```

```
    <%
        String reg=request.getParameter("reg");
        if(reg!=null){
            if(reg.equals("apply")){
    %>
    <form method="post" name="adduser" action="AddUserServlet">
    <table width="100%" vlign="center" border="1" bordercolorlight="#7777ff"
bordercolordark="#7777ff" style="border-collapse:collapse">
        <tr>
            <td colspan="2" height="25" align="center" background="images/skin/1/bg_td.gif">
            <b>
            <font color="#ffffff">新用户注册</font>
            </b>
            </td>
        </tr>
        <tr bgcolor="#dff2ed">
            <td width="40%" height="35"><b>用户名:</b><br>
                注册用户名长度限制为0-50字节
</td>
<td> <input type="text" name="user_name" size="24" maxlength="50"> 
    <input name="Submit3" type="button" value="检测用户名" onClick=CheckUser()>
</td>
</tr>
 <tr>
    <td width="40%" height="35"><b>性 别:</b><br>请选择您的性别</td>
    <td>
        <input type="radio" name="user_sex" value="男" checked>
        <img src="images/male.gif">帅哥
        <input type="radio" name="user_sex" value="女"><img src="images/female.gif">美女
  </td>
 </tr>
 <tr bgcolor="#dff2ed">
    <td width="40%" height="35"><b>密 码:(至少六位)</b>
        <br>请输入密码,区分大小写.请不要使用任何类似'*'、''或 HTML 字符
  </td>
  <td>
        <input type="password" name="user_password" size="26" maxlength="50"></td>
        </tr>
 <tr>
    <td width="40%" height="35"><b>确认密码:(至少六位)</b><br>
        请再次输入密码以便确认
  </td>
  <td>
   <input type="password" name="user_password2" size="26" onchange=CheckPass()></td>
 </tr>
```

```
<tr bgcolor="#dff2ed">
    <td width="40%" height="35"><b>密码问题：</b><br>
    请输入忘记密码的提示问题
    </td>
    <td>
      <input type="text" name="user_password_q" size="24">
    </td>
</tr>
<tr>
    <td width="40%" height="35"><b>问题答案：</b><br>
    忘记密码的提示问题答案,用于取回论坛密码
    </td>
  <td>
   <input type="text" name="user_password_a" size="24"></td>
</tr>
<tr bgcolor="#dff2ed">
 <td width="40%" height="35"><b>Email地址：</b><br>
  请输入有效的邮件地址,这将使您能用到论坛中的所有功能
   </td>
   <td>
   <input type="text" name="user_Email" size="24"></td>
 </tr>
 <tr>
 <td colspan="2" height="25" align="center" background="images/skin/1/bg_td.gif">
  <input type="submit" value="注册" onclick=CheckReg()>
  <input type="reset" value="重填"></td>
 </tr>
</table>
</form>
<%
    }}else{
 %>
 <form method="post" action="reg.jsp">
      <table width="100%" border="1" bordercolorlight="#7777ff" bordercolordark=
"#7777ff" style="border-collapse:collapse">
          <tr>
              <td height="25" align="center" background="images/skin/1/bg_td.gif"><b>
                <font color="#ffffff">服务条款和声明</font></b>
            </td>
          </tr>
          <tr bgcolor="#dff2ed">
              <td height="80" align="center">你没有接受论坛注册的《服务条款和声明》,
            请返回论坛声明页面接受方能进行注册.<p>点击下面按钮可返回注册声明页面
            </td>
          </tr>
```

```
            <tr>
                <td height="25" align="center" background="images/skin/1/bg_td.gif">
                    <input type="submit" value="返 回">
                </td>
            </tr>
        </table>
        </form>
        <%}%>
    </body>
</html>
```

代码模板 regsucess.jsp 如下，效果如图 12-6 所示。

图 12-6　注册提示页面

```
<%@page pageEncoding="gb2312"%>
<%@page contentType="text/html;charset=gb2312" %>
<%@include file="top.jsp" %>
<html>
 <head>
    <title>冰鱼论坛----注册成功</title>
    <script Language="Javascript" src="include/form.js"></Script>
 </head>
 <body style="text-align: center">
    <div align=left>>> 欢迎光临 <b>冰鱼论坛</b></div>
     < table width = " 100%" height = " 30 " border = " 0 " bordercolorlight = " # 7777ff "
bordercolordark="#7777ff" style="border-collapse:collapse">
        <tr bgcolor="#dff2ed">
            <td><img src="images/skin/1/forum_nav.gif">
                <a class=zh href="">冰鱼论坛</a> →
                <a class=zh href="reg.jsp">论坛注册</a> → 注册成功
            </td>
        </tr>
    </table>
    <%
```

```
        String reg=request.getParameter("reg");
        if(reg!=null){
            if(reg.equals("ok")){
    %>
    <form method="post" name="adduser" action="index.jsp">
    <table width="100%" border="1" bordercolorlight="#7777ff" bordercolordark=
"#7777ff" style="border-collapse:collapse">
        <tr>
            <td height="25" align="center" background="images/skin/1/bg_td.gif"><b>
                <font color="#ffffff">注册成功：冰鱼论坛欢迎您的到来!</font></b>
            </td>
        </tr>
        <tr bgcolor="#dff2ed">
            <td height="80" align="center"><a class=zh href="index.jsp">
                <li>恭喜你!注册成功,请点击这里返回论坛首页进行登录.</li></a>
            </td>
        </tr>
        <tr>
            <td colspan="2" height="25" align="center" background="images/skin/1/bg_td.gif">
            </td>
        </tr>
    </table>
    <form>
    <%
        }}else{
    %>
    <form method="post" action="reg.jsp">
    <table width="100%" border="1" bordercolorlight="#7777ff" bordercolordark=
"#7777ff" style="border-collapse:collapse">
        <tr>
            <td height="25" align="center" background="images/skin/1/bg_td.gif"><b>
                <font color="#ffffff">注册失败!</font></b>
            </td>
        </tr>
        <tr bgcolor="#dff2ed">
            <td height="80" align="center">
                你没有成功注册,原因可能是你注册的用户名已存在,
                也可能是你没有正确地提交注册信息.
                <p>点击下面按钮返回注册声明页面
            </td>
        </tr>
        <tr>
            <td height="25" align="center" background="images/skin/1/bg_td.gif">
                <input type="submit" value="返 回">
            </td>
```

```
            </tr>
        </table>
        </form>
        <%}%>
    </body>
</html>
```

2. 控制器(Servlet)

用户注册模块所涉及到的控制器 Servlet 是 AddUserServlet 和 CheckUser(见 12.3.4 节的 web.xml 配置文件)。AddUserServlet 控制器获取视图的请求后,将视图中的信息封装在实体模型 UserBean(见 12.4.2 节)中。如果获取的请求是检查用户名是否可用,则调用 CheckUser 中的方法来进行判断,并弹出提示对话框。不管用户名是否存在,都会弹出对应的提示窗口,如图 12-7 所示。

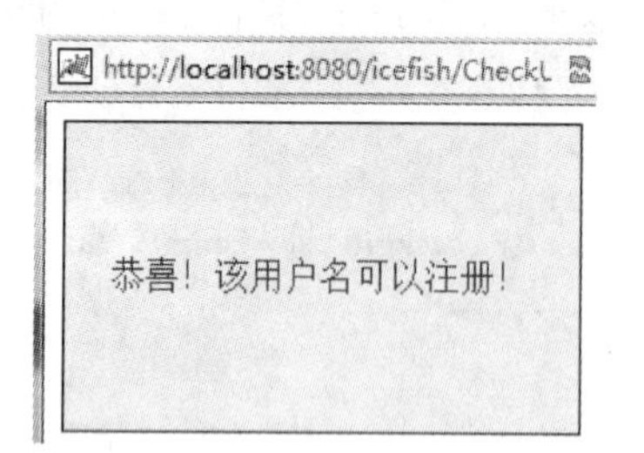

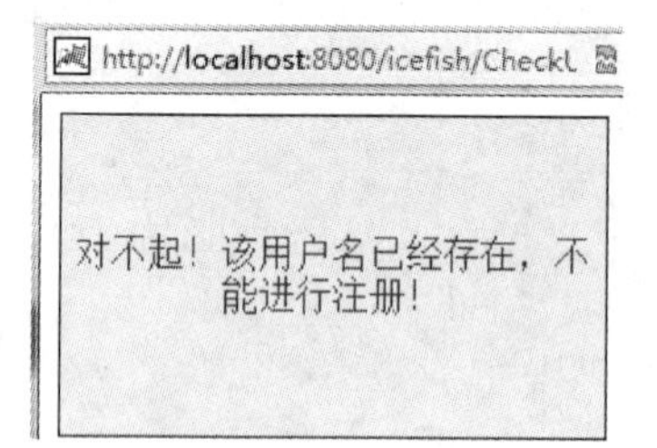

图 12-7　检测操作反馈信息

代码模板 AddUserServlet.java 如下:

```
public class AddUserServlet extends HttpServlet {
  public void init(ServletConfig config) throws ServletException {
      super.init(config);
  }
  public void destroy() {
      super.destroy();
  }
  protected void doGet(HttpServletRequest request,
          HttpServletResponse response) throws ServletException, IOException {
              doPost(request, response);
  }
  protected void doPost(HttpServletRequest request,
      HttpServletResponse response) throws ServletException, IOException {
      response.setContentType("text/html;charset=GB2312");
      request.setCharacterEncoding("GB2312");
      PrintWriter out=response.getWriter();
      HttpSession session=request.getSession(true);

      String user_name=request.getParameter("user_name");
      String user_password=request.getParameter("user_password");
```

```
    String user_sex=request.getParameter("user_sex");
    String user_password_q=request.getParameter("user_password_q");
    String user_password_a=request.getParameter("user_password_a");
    String user_Email=request.getParameter("user_Email");

    UserDataBean udb=new UserDataBean();
    UserBean userBean=new UserBean();
    userBean.setUser_Name(user_name);
    userBean.setUser_Password(user_password);
    userBean.setUser_Password_a(user_password_a);
    userBean.setUser_Password_q(user_password_q);
    userBean.setUser_Email(user_Email);
    userBean.setUser_Sex(user_sex);
    boolean checkUser=udb.checkUser(userBean);
    if(checkUser){
     boolean result = udb.addUser(userBean);
     RequestDispatcher rd = request.getRequestDispatcher("regsuccess.jsp?reg=ok");
     rd.forward(request, response);
        }
    else{
     RequestDispatcher rd = request.getRequestDispatcher("regsuccess.jsp");
     rd.forward(request, response);
        }
    }else{
        RequestDispatcher rd = request.getRequestDispatcher("regsuccess.jsp");
        rd.forward(request, response);
    }
  }
}
```

代码模板 CheckUser.java 如下：

```
public class CheckUser extends HttpServlet {
 public void init(ServletConfig config) throws ServletException {
     super.init(config);
 }
 public void destroy() {
     super.destroy();
 }
 protected void doGet(HttpServletRequest request,
         HttpServletResponse response) throws ServletException, IOException {
     doPost(request, response);
 }
 protected void doPost(HttpServletRequest request,
         HttpServletResponse response) throws ServletException, IOException {
     response.setContentType("text/html;charset=GB2312");
     request.setCharacterEncoding("GB2312");
     PrintWriter out=response.getWriter();
     HttpSession session=request.getSession(true);
```

```
        String user_name=request.getParameter("user_name");
        UserDataBean udb=new UserDataBean();
        UserBean userBean=new UserBean();
        userBean.setUser_Name(user_name);
        boolean checkUser=udb.checkUser(userBean);
        if(checkUser){
            out.println("<table width=100% height=100% align=center vlign=center border=1
                    bordercolorlight=#7777ff bordercolordark=#7777ff style=border-collapse:collapse>"
                    +"<tr bgcolor=#dff2ed align=center><td><font color=blue>
                    恭喜!该用户名可以注册!</font></td></td>"+"</table>");
        }else{
            out.println("<table width=100% height=100% align=center vlign=center border=1
                    bordercolorlight=#7777ff bordercolordark=#7777ff style=border-collapse:collapse>"
                    +"<tr bgcolor=#dff2ed align=center><td><font color=blue>
                    对不起!该用户名已经存在,不能进行注册!</font></td></td>"+"</table>");
        }
    }
}
```

如果获取的请求是检查密码等信息是否为空，则调用 form.js(JavaScript 页面)中的方法进行检测。form.js 针对该功能的代码如下(由于 form.js 还有其他功能，所以在此并未列出其全部代码)：

```
//注册页面检测两次密码是否输入相同
function CheckPass()
{
 if(document.adduser.user_password.value==''||document.adduser.user_password2.value=='')
 {
      alert("你还没有输入密码,请输入密码!");
      window.location.reload();
 }elseif(document.adduser.user_password.value!=document.adduser.user_password2.value){
      alert("两次输入的密码不一样,请重新输入!");
      window.location.reload();
 }
}
//检测用户名是否已存在
function CheckUser()
{
 if(document.adduser.user_name.value==''){
      alert("你还没有输入用户名,请输入用户名!");
 }else{
      window.open("CheckUser?user_name="+document.adduser.user_name.value,"","width=
250,height=150");
 }
}
//提交注册信息时的验证
function CheckReg()
{
```

```
 if(document.adduser.user_name.value=='' || document.adduser.user_password.value=='' ||
document.adduser.user_password2.value==''){
      alert("你还没有输入用户名或者密码,请输入用户名或者密码!");
      window.location.reload();
 }else if(document.adduser.user_password.value!=document.adduser.user_password2.value){
      alert("两次输入的密码不一样,请重新输入!");
      window.location.reload();
 }else if(document.adduser.user_password_a.value=='' || document.adduser.user_password_q
.value==''){
      alert("请填入密码问题和密码答案,方便以后取回论坛密码!");
      window.location.reload();
 }
}
```

如果获取的请求是注册,AddUserServlet 则调用 UserDataBean(见 12.4.3 节)业务模型中的 adduser()等方法进行注册。注册完成后进入 regsuccess.jsp 页面,提示用户注册成功或注册失败,显示效果如前面的图 12-5 所示。如果成功注册,则转到 index.jsp 页面,可以登录系统;如果失败,则转到 reg.jsp 页面,继续重新注册。

12.5.2 用户登录

用户输入自己的用户名和密码以及验证码之后,论坛系统将对用户名和密码进行验证。如果用户名和密码以及验证码同时正确,则登录成功,进入系统主界面;如果用户名、密码或者验证码有误,则回到登录页面继续重新登录。

1. 视图(JSP)

页面 login.jsp 提供登录信息输入界面,效果如图 12-8 所示。

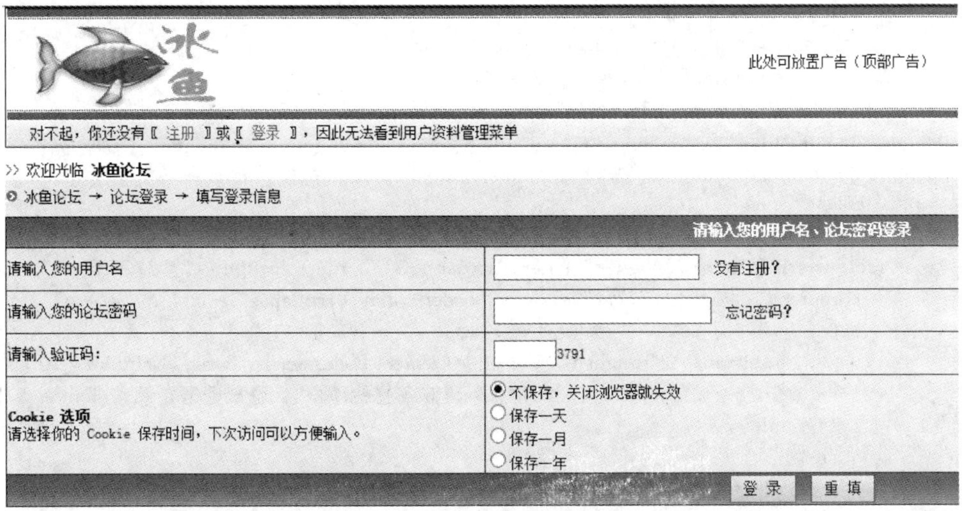

图 12-8　登录界面

代码模板 login.jsp 如下：

```
<%@page pageEncoding="gb2312"%>
<%@page contentType="text/html;charset=gb2312" %>
<%@include file="top.jsp" %>
<%@page import="java.util.*" %>
<html>
 <head>
     <title>冰鱼论坛----论坛注册新用户</title>
     <script Language="Javascript" src="include/form.js"></Script>
 </head>
<%
// 取随机产生的认证码(4位数字)
String sRand="";
//生成随机类
Random random = new Random();
for (int i=0;i<4;i++){
 String rand=String.valueOf(random.nextInt(10));
 sRand+=rand;
}
// 将认证码存入 SESSION
session.setAttribute("rand",sRand);
%>
 <body style="text-align: center">
     <div align=left>>> 欢迎光临 <b>冰鱼论坛</b></div>
      < table width =" 100%" height =" 30" border =" 0" bordercolorlight =" #7777ff"
bordercolordark="#7777ff"
          style="border-collapse:collapse">
          <tr bgcolor="#dff2ed">
               <td><img src="images/skin/1/forum_nav.gif">
                    <a class=zh href="">冰鱼论坛</a> →
                    <a class=zh href="login.jsp">论坛登录</a> → 填写登录信息
               </td>
          </tr>
     </table>
     <form method="post" name="userlogin" action="check.jsp">
     <table width="100%" vlign="center" border="1" bordercolorlight="#7777ff"
          bordercolordark="#7777ff" style="border-collapse:collapse">
          <tr>
               <td colspan="2" height="25" align="center" background="images/skin/1/bg_td.gif">
                    <b><font color="#ffffff">请输入您的用户名、论坛密码登录</font></b>
               </td>
          </tr>
          <tr bgcolor="#dff2ed">
               <td width="30%" height="35">请输入您的用户名</td>
               <td><input type="text" name="user_name" size="24">
```

```
                <a class=zh href="reg.jsp">没有注册?</a>
            </td>
        </tr>
        <tr bgcolor="#dff2ed">
            <td width="30%" height="35">请输入您的论坛密码</td>
            <td><input type="password" name="user_password" size="26">
                <a class=zh href="">忘记密码?</a>
            </td>
        </tr>
        <tr bgcolor="#dff2ed">
            <td width="30%" height="35">请输入验证码: </td>
            <td><input name="checknumber" type="text" maxlength="4" size="6">
                <%out.print(sRand);%>
         </td>
        </tr>
        <tr bgcolor="#dff2ed">
            <td width="30%"><b>Cookie 选项</b>
                <br>请选择你的 Cookie 保存时间,下次访问可以方便输入.
            </td>
            <td>
                <input type="radio" name="save_login" value="no_time" checked>
                不保存,关闭浏览器就失效
                <br><input type="radio" name="save_login" value="one_day">保存一天
                <br><input type="radio" name="save_login" value="one_month">保存一月
                <br><input type="radio" name="save_login" value="one_year">保存一年
            </td>
        </tr>
        <tr>
            <td colspan="2" height="25" align="center" background="images/skin/1/bg_td.gif">
                <input type="submit" onclick="CheckLogin()" value="登 录">
                <input type="reset" value="重 填">
            </td>
        </tr>
    </table>
    </form>
  </body>
</html>
```

2. 控制器(Servlet)

用户登录模块所涉及的控制器 Servlet 对象的名称是 LoginServlet(见 12.3.4 节中的 web.xml 配置文件)。控制器获取视图的请求后,将视图中的信息封装在实体模型 User(见 12.4.2 节)中,然后调用业务模型 UserDataBean(见 12.4.3 节)中的登录方法执行相应的业务处理。登录成功进入系统主界面(如图 12-9 所示),登录失败则返回 login.jsp 页面。

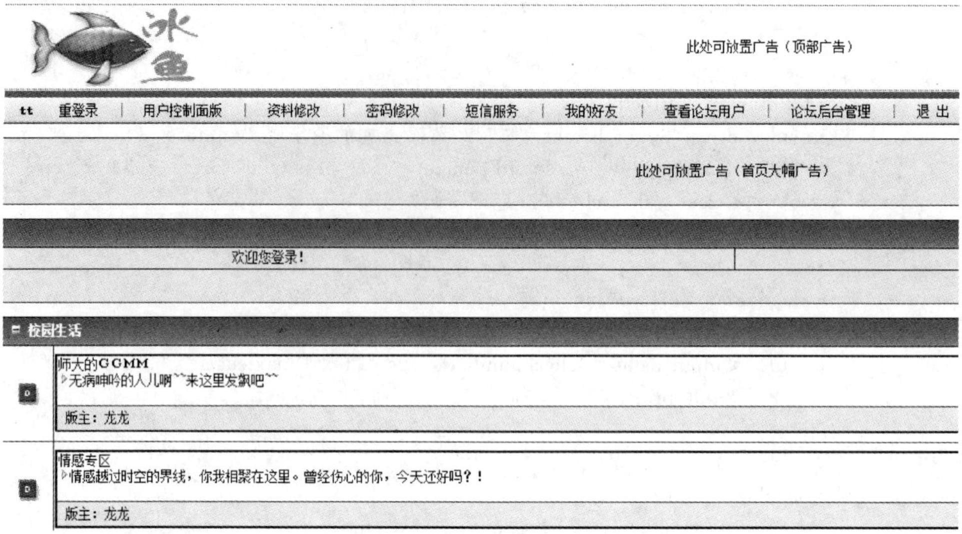

此处可放置广告（顶部广告）

tt 重登录 | 用户控制面版 | 资料修改 | 密码修改 | 短信服务 | 我的好友 | 查看论坛用户 | 论坛后台管理 | 退出

此处可放置广告（首页大幅广告）

欢迎您登录！

校园生活

师大的GGMM
无病呻吟的人儿啊~~来这里发飙吧~~
版主：龙龙

情感专区
情感越过时空的界线，你我相聚在这里。曾经伤心的你，今天还好吗？！
版主：龙龙

图 12-9 系统主界面

代码模板 LoginServlet.java 如下：

```
package net.icefish.servlet;
import net.icefish.bean.*;
import java.net.*;
import java.io.IOException;
import java.io.PrintWriter;
import javax.servlet.*;
import javax.servlet.http.*;
public class LoginServlet extends HttpServlet {
 public void init(ServletConfig config) throws ServletException {
      super.init(config);
}
 public void destroy() {
      super.destroy();
 }
 protected void doGet(HttpServletRequest request,
          HttpServletResponse response) throws ServletException, IOException {
      doPost(request, response);
 }
 protected void doPost(HttpServletRequest request,
          HttpServletResponse response) throws ServletException, IOException {
      response.setContentType("text/html;charset=GB2312");
      request.setCharacterEncoding("GB2312");
      PrintWriter out=response.getWriter();
      HttpSession session=request.getSession(true);
```

```
String save_login=request.getParameter("save_login");
String user_name=request.getParameter("user_name");
String user_password=request.getParameter("user_password");
//用户登录验证
UserDataBean udb=new UserDataBean();
UserBean userBean=new UserBean();
userBean.setUser_Name(user_name);
userBean.setUser_Password(user_password);
boolean result=udb.loginUser(userBean);
if(result){
    //处理 cookie
    String str_user_name = URLEncoder.encode(user_name);
    Cookie user_name_cookie = new Cookie("icefish_user_name", str_user_name);
    Cookie user_password_cookie = new Cookie("icefish_user_password", user_password);
  if(save_login==null){
  }
    else if(save_login.equals("no_time")){
        user_name_cookie.setMaxAge(0);
        user_password_cookie.setMaxAge(0);
        response.addCookie(user_name_cookie);
        response.addCookie(user_password_cookie);
    }
    else if(save_login.equals("one_day")){
        int oneday=60 * 60 * 24;
        user_name_cookie.setMaxAge(oneday);
        user_password_cookie.setMaxAge(oneday);
        response.addCookie(user_name_cookie);
        response.addCookie(user_password_cookie);
    }
    else if(save_login.equals("one_month")){
        int onemonth=60 * 60 * 24 * 31;
        user_name_cookie.setMaxAge(onemonth);
        user_password_cookie.setMaxAge(onemonth);
        response.addCookie(user_name_cookie);
        response.addCookie(user_password_cookie);
    }
    else if(save_login.equals("one_year")){
        int oneyear=60 * 60 * 24 * 365;
        user_name_cookie.setMaxAge(oneyear);
        user_password_cookie.setMaxAge(oneyear);
        response.addCookie(user_name_cookie);
        response.addCookie(user_password_cookie);
    }
```

```
RequestDispatcher rd = request.getRequestDispatcher("index.jsp?login=pass&user_name=
                    "+user_name+"&user_password="+user_password);
rd.forward(request, response);
}else{
    out.println("<script>alert('登录失败,用户名或密码错误!!!!');
            this.location.href='index.jsp';</script>");
}
  }
}
```

12.5.3 版块管理

当管理员用户登录成功后,可以在如图 12-8 所示的主界面中选择"论坛后台管理"菜单,进行各项管理操作,包括系统管理、版块管理、用户管理、帖子管理、友情链接管理等,如图 12-10 所示。本书主要介绍版块管理的相关实现,包括增加版块、编辑版块、删除版块。其他管理操作与版块管理类似,不逐一细述。

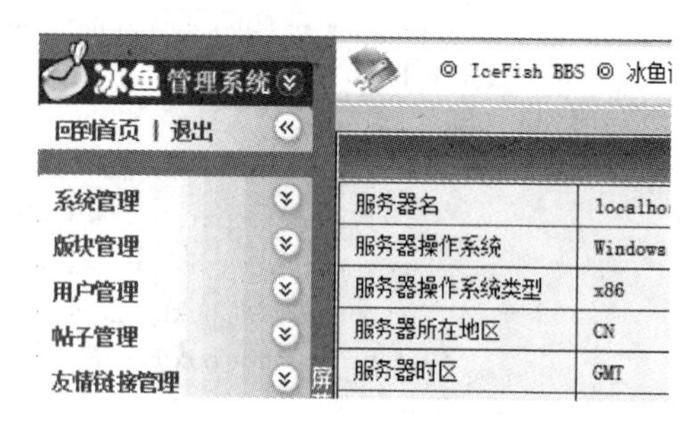

图 12-10 论坛后台管理界面

1. 视图(JSP)

页面 admin_addboard.jsp 提供添加版块的信息输入界面;admin_editboard.jsp 和 admin_delboard.jsp 分别提供编辑操作界面和删除操作界面。由于这几个页面代码主体很相似,所以在此只列出 admin_editboard.jsp 的相关代码,其运行效果如图 12-11 所示。

图 12-11 添加版块界面

将全部信息输入完毕，单击“添加版块”按钮成功添加后，就可以在论坛版块中查看到新增加的“Game”版块，并且可以在此基础上对其进行编辑版块和删除版块的操作，如图 12-12 所示。单击“编辑该版块”选项将回到图 12-11 所示的界面，可以对版块的相关信息进行再次编辑；单击“删除该版块”选项将会删除“Game”版块。

图 12-12　新增版块操作选项

代码模板 admin_addboard.jsp 如下：

```
<%@page pageEncoding="gb2312"%>
<%@page contentType="text/html;charset=gb2312" %>
<%@include file="admin_top.jsp" %>
<%@page import="java.sql.*"%>
<jsp:useBean id="indexBean" scope="page" class="net.icefish.bean.IndexBean" />
<html>
<body>
 <table cellpadding=3 cellspacing=1 align=center class="tableBorder" style="width:96%">
 <tr align=center>
     <td width="100%" height=25 colspan=2 background="../images/skin/1/bg_td.gif">
         <font color="#ffffff"><b>版块管理</b></font>
     </td>
 </tr>
 <tr>
     <td width="100%" class="forumRowHighlight" colspan=2><b>注意: </b>
     <br>①删除版块同时将删除该版块下所有帖子!删除分类同时删除下属版块和其中帖子! 操作时请完整填写表单信息.<br>②如果选择复位所有版面,则所有版面都将作为一级论坛(分类),这时您需要重新对各个版面进行归属的基本设置,不要轻易使用该功能,仅在做出了错误的设置而无法复原版面之间的关系和排序的时候使用,在这里您也可以只针对某个分类进行复位操作(见分类的更多操作下拉菜单),具体请看操作说明.
     <br><font color="#0000ff">每个版面的更多操作请见下拉菜单,操作前请仔细阅读说明,分类下拉菜单中比别的版面增加了分类排序和分类复位功能.</font>
     </td>
   </tr>
 </table>
 <form name="addboard" method="post" action="BoardServlet?action=add">
 <table cellpadding=3 cellspacing=1 align=center class="tableBorder" style="width:96%">
     <tr align=center>
         <td width="100%" height=25 colspan=2 background="../images/skin/1/bg_td.gif">
             <font color="#ffffff"><b>添加新版块</b></font>
            </td>
     </tr>
     <tr>
         <td width="100%" class="forumRowHighlight" colspan=2>说明:
```

```
        <br>1.添加论坛版面后,相关的设置均为默认设置,请返回论坛版面管理首页版面列表
的高级设置中设置该论坛的相应属性,如果您想对该论坛做更具体的权限设置,请到论坛权限管理
中设置相应用户组在该版面的权限.<br>2.<font color="#0000ff">如果您添加的是论坛分类
</font>,只需要在所属分类中选择作为论坛分类即可;
        <font color="#0000ff">如果您添加的是论坛版面</font>,
        则要在所属分类中确定并选择该论坛版面的上级版面
    </td>
    </tr>
    <tr>
        <td class="Forumrow" width="40%">论坛名称</td>
        <td class="Forumrow" width="60%">
            <input type="text" name="board_name" size="26">
        </td>
    </tr>
    <tr>
        <td class="Forumrow" width="40%">版面说明<br>可以使用 HTML 代码</td>
        <td class="Forumrow" width="60%">
            <textarea rows="4" name="board_info" cols="31"></textarea>
        </td>
    </tr>
    <tr>
        <td class="Forumrow" width="40%">所属版块</td>
        <td class="Forumrow" width="60%">
            <select name="board_bid"><option value="">论坛分类一级版块
            <%
            String sql3="select * from icefish_board WHERE board_isMother='true' order by
                    board_id asc";
            ResultSet rs3=null;
            rs3=indexBean.executeSQL(sql3);
            while(rs3.next()){%>
            <option value="<%=rs3.getString("board_id")%>">
                    <%=rs3.getString("board_name")%>
            <%
            }
            rs3.close();
            indexBean.close();%>
            </select></td>
    </tr>
    <tr>
        <td class="Forumrow" width="40%">论坛版主
            <br>请注意填写正确的版主名称,否则所设的版主将无效
        </td>
        <td class="Forumrow" width="60%">
            <input type="text" name="board_master" size="26">
        </td>
```

```
    </tr>
    <tr>
        <td class="Forumrow" width="40%"></td>
        <td class="Forumrow" width="60%">
            <input type="submit" value="添加版块" onclick="AddBoard()">
        </td>
    </tr>
  </table>
  </form>
</body>
<%@include file="admin_foot.jsp" %>
</html>
```

2. 控制器(Servlet)

该控制器 Servlet 对象的名称是 BoardServlet(见 12.3.4 节的 web.xml 配置文件)。控制器获取视图的请求后,将视图中的信息封装在实体模型 BoardBean(见 12.4.2 节)中,然后调用业务模型 BoardDataBean(见 12.4.3 节)中的方法执行添加、修改和删除的业务处理。

代码模板 BoardServlet.java 如下:

```
package net.icefish.servlet;
import net.icefish.bean.*;
import java.io.IOException;
import java.io.PrintWriter;
import javax.servlet.*;
import javax.servlet.http.*;
public class BoardServlet extends HttpServlet {
 public void init(ServletConfig config) throws ServletException {
     super.init(config);
 }
 public void destroy() {
     super.destroy();
 }
 protected void doGet(HttpServletRequest request,
         HttpServletResponse response) throws ServletException, IOException {
     doPost(request, response);
 }

 protected void doPost(HttpServletRequest request,
         HttpServletResponse response) throws ServletException, IOException {
     response.setContentType("text/html;charset=GB2312");
     request.setCharacterEncoding("GB2312");
     PrintWriter out=response.getWriter();
     String action=request.getParameter("action");
```

```
    BoardDataBean bdb=new BoardDataBean();
    BoardBean bb=new BoardBean();
    if(action.equals("add")){
        String board_name=request.getParameter("board_name");
        String board_info=request.getParameter("board_info");
        String board_master=request.getParameter("board_master");
        String board_bid=request.getParameter("board_bid");
        if(board_bid.equals("")){
            bb.setBoard_Name(board_name);
            bb.setBoard_Info(board_info);
            bb.setBoard_Master(board_master);
            bb.setBoard_IsMother(true);
            boolean result=bdb.addBoard(bb);
            if(result){
                out.println("<script>alert('添加新版块成功!');
                        location.href='admin_board.jsp';</script>");
            }else{
                out.println("<script>alert('添加新版块失败!请输入正确的信息再点击添加.');
                        window.location.reload();</script>");
            }
        }else{
            bb.setBoard_Name(board_name);
            bb.setBoard_Info(board_info);
            bb.setBoard_Master(board_master);
            bb.setBoard_IsMother(false);
            bb.setBoard_BID(board_bid);
            boolean result=bdb.addBoard(bb);
            if(result){
                out.println("<script>alert('添加新版块成功!');
                        location.href='admin_board.jsp';</script>");
            }else{
                out.println("<script>alert('添加新版块失败!请输入正确的信息再点击添加.');
                        window.location.reload();</script>");
            }
        }

    }
    else if(action.equals("edit")){
        String board_id=request.getParameter("board_id");
        String board_name=request.getParameter("board_name");
        String board_info=request.getParameter("board_info");
        String board_master=request.getParameter("board_master");
        String board_bid=request.getParameter("board_bid");
        if(board_bid.equals("")){
            bb.setBoard_ID(board_id);
```

```
            bb.setBoard_Name(board_name);
            bb.setBoard_Info(board_info);
            bb.setBoard_Master(board_master);
            bb.setBoard_IsMother(true);
            boolean result=bdb.editBoard(bb);
            if(result){
                out.println("<script>alert('修改版块成功!');
                        location.href='admin_board.jsp';</script>");
            }else{
                out.println("<script>alert('修改版块失败!请输入正确的信息再点击添加.');
                        window.location.reload();</script>");
            }
        }else{
            bb.setBoard_ID(board_id);
            bb.setBoard_Name(board_name);
            bb.setBoard_Info(board_info);
            bb.setBoard_Master(board_master);
            bb.setBoard_IsMother(false);
            bb.setBoard_BID(board_bid);
            boolean result=bdb.editBoard(bb);
            if(result){
                out.println("<script>alert('修改版块成功!');
                        location.href='admin_board.jsp';</script>");
            }else{
                out.println("<script>alert('修改版块失败!请输入正确的信息再点击添加.');
                        window.location.reload();</script>");
            }
        }
    }
    else if(action.equals("del")){
        String board_id=request.getParameter("board_id");
        String board_isMother=request.getParameter("board_isMother");
        if(board_isMother.equals("true")){
            bb.setBoard_IsMother(true);
            bb.setBoard_ID(board_id);
            boolean result=bdb.delBoard(bb);
            if(result){
                out.println("<script>alert('删除版块成功!');
                        location.href='admin_board.jsp';</script>");
            }else{
                out.println("<script>alert('删除版块失败!请再试一次.');
                        window.location.reload();</script>");
            }
        }else{
            bb.setBoard_IsMother(false);
```

```
                    bb.setBoard_ID(board_id);
                    boolean result=bdb.delBoard(bb);
                    if(result){
                        out.println("<script>alert('删除版块成功!');
                                location.href='admin_board.jsp';</script>");
                    }else{
                        out.println("<script>alert('删除版块失败!请再试一次.');
                                window.location.reload();</script>");
                    }
                }
            }
        }
    }
```

附录 A

HTML 常用标签

表 A-1 是 HTML 常用标签一览表。

表 A-1　HTML 标签一览表

标　签	描　述
<!——…——>	定义注释
<! DOCTYPE>	定义文档类型
<a>	定义锚
<abbr>	定义缩写
<acronym>	定义只取首字母的缩写
<address>	定义文档作者或拥有者的联系信息
<applet>	不赞成使用。定义嵌入的 applet
<area>	定义图像映射内部的区域
<article>	定义文章
<aside>	定义页面内容之外的内容
<audio>	定义声音内容
<b>	定义粗体字
<base>	定义页面中所有链接的默认地址或默认目标
<basefont>	不赞成使用。定义页面中文本的默认字体、颜色或尺寸
<bdi>	定义文本的文本方向，使其脱离其周围文本的方向设置
<bdo>	定义文字方向
<big>	定义大号文本
<blockquote>	定义长的引用
<body>	定义文档的主体
 	定义简单的折行
<button>	定义按钮（push button）
<canvas>	定义图形
<caption>	定义表格标题
<center>	不赞成使用。定义居中文本
<cite>	定义引用(citation)
<code>	定义计算机代码文本
<col>	定义表格中一个或多个列的属性值
<colgroup>	定义表格中供格式化的列组

续表

标　签	描　述
<command>	定义命令按钮
<datalist>	定义下拉列表
<dd>	定义列表中项目的描述
<del>	定义被删除文本
<details>	定义元素的细节
<dir>	不赞成使用。定义目录列表
<div>	定义文档中的节
<dfn>	定义项目
<dialog>	定义对话框或窗口
<dl>	定义列表
<dt>	定义列表中的项目
<em>	定义强调文本
<embed>	定义外部交互内容或插件
<fieldset>	定义围绕表单中元素的边框
<figcaption>	定义 figure 元素的标题
<figure>	定义媒介内容的分组，以及它们的标题
<font>	不赞成使用。定义文字的字体、尺寸和颜色
<footer>	定义 section 或 page 的页脚
<form>	定义供用户输入的 HTML 表单
<frame>	定义框架集的窗口或框架
<frameset>	定义框架集
<h1> to <h6>	定义 HTML 标题
<head>	定义关于文档的信息
<header>	定义 section 或 page 的页眉
<hr>	定义水平线
<html>	定义 HTML 文档
<i>	定义斜体字
<iframe>	定义内联框架
<img>	定义图像
<input>	定义输入控件
<ins>	定义被插入文本
<isindex>	不赞成使用。定义与文档相关的可搜索索引
<kbd>	定义键盘文本
<keygen>	定义生成密钥
<label>	定义 input 元素的标注
<legend>	定义 fieldset 元素的标题
<li>	定义列表的项目

续表

标　签	描　述
<link>	定义文档与外部资源的关系
<map>	定义图像映射
<mark>	定义有记号的文本
<menu>	定义命令的列表或菜单
<menuitem>	定义用户可以从弹出菜单调用的命令/菜单项目
<meta>	定义关于 HTML 文档的元信息
<meter>	定义预定义范围内的度量
<nav>	定义导航链接
<noframes>	定义针对不支持框架的用户的替代内容
<noscript>	定义针对不支持客户端脚本的用户的替代内容
<object>	定义内嵌对象
<ol>	定义有序列表
<optgroup>	定义选择列表中相关选项的组合
<option>	定义选择列表中的选项
<output>	定义输出的一些类型
<p>	定义段落
<param>	定义对象的参数
<pre>	定义预格式文本
<progress>	定义任何类型的任务的进度
<q>	定义短的引用
<rp>	定义若浏览器不支持 ruby 元素显示的内容
<rt>	定义 ruby 注释的解释
<ruby>	定义 ruby 注释
<s>	不赞成使用。定义加删除线的文本
<samp>	定义计算机代码样本
<script>	定义客户端脚本
<section>	定义 section
<select>	定义选择列表(下拉列表)
<small>	定义小号文本
<source>	定义媒介源
<span>	定义文档中的节
<strike>	不赞成使用。定义加删除线文本
<strong>	定义强调文本
<style>	定义文档的样式信息
<sub>	定义下标文本
<summary>	为 <details> 元素定义可见的标题
<sup>	定义上标文本

续表

标　　签	描　　述
<table>	定义表格
<tbody>	定义表格中的主体内容
<td>	定义表格中的单元
<textarea>	定义多行的文本输入控件
<tfoot>	定义表格中的表注内容(脚注)
<th>	定义表格中的表头单元格
<thead>	定义表格中的表头内容
<time>	定义日期/时间
<title>	定义文档的标题
<tr>	定义表格中的行
<track>	定义用在媒体播放器中的文本轨道
<tt>	定义打字机文本
<u>	不赞成使用。定义下划线文本
<ul>	定义无序列表
<var>	定义文本的变量部分
<video>	定义视频
<wbr>	定义转换行
<xmp>	不赞成使用。定义预格式文本

附录 B

HTML 中的颜色表示

表 B-1 是 HTML 中使用的颜色值。

表 B-1　HTML 颜色列表

颜　色　名	十六进制颜色值
AliceBlue	#F0F8FF
AntiqueWhite	#FAEBD7
Aqua	#00FFFF
Aquamarine	#7FFFD4
Azure	#F0FFFF
Beige	#F5F5DC
Bisque	#FFE4C4
Black	#000000
BlanchedAlmond	#FFEBCD
Blue	#0000FF
BlueViolet	#8A2BE2
Brown	#A52A2A
BurlyWood	#DEB887
CadetBlue	#5F9EA0
Chartreuse	#7FFF00
Chocolate	#D2691E
Coral	#FF7F50
CornflowerBlue	#6495ED
Cornsilk	#FFF8DC
Crimson	#DC143C
Cyan	#00FFFF
DarkBlue	#00008B
DarkCyan	#008B8B
DarkGoldenRod	#B8860B
DarkGray	#A9A9A9
DarkGreen	#006400
DarkKhaki	#BDB76B
DarkMagenta	#8B008B

续表

颜　色　名	十六进制颜色值
DarkOliveGreen	#556B2F
Darkorange	#FF8C00
DarkOrchid	#9932CC
DarkRed	#8B0000
DarkSalmon	#E9967A
DarkSeaGreen	#8FBC8F
DarkSlateBlue	#483D8B
DarkSlateGray	#2F4F4F
DarkTurquoise	#00CED1
DarkViolet	#9400D3
DeepPink	#FF1493
DeepSkyBlue	#00BFFF
DimGray	#696969
DodgerBlue	#1E90FF
Feldspar	#D19275
FireBrick	#B22222
FloralWhite	#FFFAF0
ForestGreen	#228B22
Fuchsia	#FF00FF
Gainsboro	#DCDCDC
GhostWhite	#F8F8FF
Gold	#FFD700
GoldenRod	#DAA520
Gray	#808080
Green	#008000
GreenYellow	#ADFF2F
HoneyDew	#F0FFF0
HotPink	#FF69B4
IndianRed	#CD5C5C
Indigo	#4B0082
Ivory	#FFFFF0
Khaki	#F0E68C
Lavender	#E6E6FA
LavenderBlush	#FFF0F5
LawnGreen	#7CFC00
LemonChiffon	#FFFACD
LightBlue	#ADD8E6

续表

颜　色　名	十六进制颜色值
LightCoral	#F08080
LightCyan	#E0FFFF
LightGoldenRodYellow	#FAFAD2
LightGrey	#D3D3D3
LightGreen	#90EE90
LightPink	#FFB6C1
LightSalmon	#FFA07A
LightSeaGreen	#20B2AA
LightSkyBlue	#87CEFA
LightSlateBlue	#8470FF
LightSlateGray	#778899
LightSteelBlue	#B0C4DE
LightYellow	#FFFFE0
Lime	#00FF00
LimeGreen	#32CD32
Linen	#FAF0E6
Magenta	#FF00FF
Maroon	#800000
MediumAquaMarine	#66CDAA
MediumBlue	#0000CD
MediumOrchid	#BA55D3
MediumPurple	#9370D8
MediumSeaGreen	#3CB371
MediumSlateBlue	#7B68EE
MediumSpringGreen	#00FA9A
MediumTurquoise	#48D1CC
MediumVioletRed	#C71585
MidnightBlue	#191970
MintCream	#F5FFFA
MistyRose	#FFE4E1
Moccasin	#FFE4B5
NavajoWhite	#FFDEAD
Navy	#000080
OldLace	#FDF5E6
Olive	#808000
OliveDrab	#6B8E23
Orange	#FFA500
OrangeRed	#FF4500
Orchid	#DA70D6

续表

颜　色　名	十六进制颜色值
PaleGoldenRod	#EEE8AA
PaleGreen	#98FB98
PaleTurquoise	#AFEEEE
PaleVioletRed	#D87093
PapayaWhip	#FFEFD5
PeachPuff	#FFDAB9
Peru	#CD853F
Pink	#FFC0CB
Plum	#DDA0DD
PowderBlue	#B0E0E6
Purple	#800080
Red	#FF0000
RosyBrown	#BC8F8F
RoyalBlue	#4169E1
SaddleBrown	#8B4513
Salmon	#FA8072
SandyBrown	#F4A460
SeaGreen	#2E8B57
SeaShell	#FFF5EE
Sienna	#A0522D
Silver	#C0C0C0
SkyBlue	#87CEEB
SlateBlue	#6A5ACD
SlateGray	#708090
Snow	#FFFAFA
SpringGreen	#00FF7F
SteelBlue	#4682B4
Tan	#D2B48C
Teal	#008080
Thistle	#D8BFD8
Tomato	#FF6347
Turquoise	#40E0D0
Violet	#EE82EE
VioletRed	#D02090
Wheat	#F5DEB3
White	#FFFFFF
WhiteSmoke	#F5F5F5
Yellow	#FFFF00
YellowGreen	#9ACD32

附录 C

JSP 内置对象及其常用方法

1. request 对象

客户端的请求信息被封装在 request 对象中，通过它才能了解到客户的需求，然后做出响应。它是 HttpServletRequest 类的实例。其方法说明如下：

(1) object getAttribute(String name) 返回指定属性的属性值。

(2) Enumeration getAttributeNames() 返回所有可用属性名的枚举。

(3) String getCharacterEncoding() 返回字符编码方。

(4) int getContentLength() 返回请求体的长度(以字节数)。

(5) String getContentType() 得到请求体的 MIME 类型。

(6) ServletInputStream getInputStream() 得到请求体中一行的二进制流。

(7) String getParameter(String name) 返回 name 指定参数的参数值。

(8) Enumeration getParameterNames() 返回可用参数名的枚举。

(9) String[] getParameterValues(String name) 返回包含参数 name 的所有值的数组。

(10) String getProtocol() 返回请求用的协议类型及版本号。

(11) String getScheme() 返回请求用的计划名，如 http、https 及 ftp 等。

(12) String getServerName() 返回接受请求的服务器主机名。

(13) int getServerPort() 返回服务器接受此请求所用的端口号。

(14) BufferedReader getReader() 返回解码过的请求体。

(15) String getRemoteAddr() 返回发送此请求的客户端 IP 地址。

(16) String getRemoteHost() 返回发送此请求的客户端主机名。

(17) void setAttribute(String key,Object obj) 设置属性的属性值。

(18) String getRealPath(String path) 返回一个虚拟路径的真实路径。

(19) void setCharacterEncoding(String code)设置统一字符编码。

2. response 对象

response 对象包含了响应客户请求的有关信息，但在 JSP 中很少直接用到它。它是 HttpServletResponse 类的实例。其方法说明如下：

(1) String getCharacterEncoding() 返回响应用的是何种字符编码。

(2) ServletOutputStream getOutputStream() 返回响应的一个二进制输出流。

(3) PrintWriter getWriter() 返回可以向客户端输出字符的一个对象。

(4) void setContentLength(int len) 设置响应头长度。

(5) void setContentType(String type) 设置响应的 MIME 类型。

(6) void setHeader(String name,String value) 设置指定名字的 HTTP 文件头的值，以此来操作 HTTP 文件头。

(7) sendRedirect(java. lang. String location) 重新定向客户端的请求。

3. session 对象

session 对象是 HttpSession 类的实例，指的是客户端与服务器的一次会话，从客户连到服务器的一个 WebApplication 开始，直到客户端与服务器断开连接为止。其方法说明如下：

(1) long getCreationTime() 返回 session 创建时间。

(2) public String getId() 返回 session 创建时 JSP 引擎为它设的唯一 ID 号。

(3) long getLastAccessedTime() 返回此 session 里客户端最近一次请求时间。

(4) int getMaxInactiveInterval() 返回两次请求间隔多长时间此 session 被取消(ms)。

(5) String[] getValueNames() 返回一个包含此 session 中所有可用属性的数组。

(6) void invalidate() 取消 session，使 session 不可用。

(7) boolean isNew() 返回服务器创建的一个 session，客户端是否已经加入。

(8) void removeValue(String name) 删除 session 中指定的属性。

(9) void setMaxInactiveInterval() 设置两次请求间隔多长时间此 session 被取消(ms)。

(10) getAttribute(String name) 获取 session 属性值。

(11) setAttribute(String name) 设置 session 属性值。

4. Out 对象

out 对象是 JspWriter 类的实例，是向客户端输出内容常用的对象。其方法说明如下：

(1) void clear() 清除缓冲区的内容。

(2) void clearBuffer() 清除缓冲区的当前内容。

(3) void flush() 清空流。

(4) int getBufferSize() 返回缓冲区以字节数的大小，如不设缓冲区则为 0。

(5) int getRemaining() 返回缓冲区还剩余多少可用。

(6) boolean isAutoFlush() 返回缓冲区满时，是自动清空还是抛出异常。

(7) void close() 关闭输出流。

5. page 对象

page 对象就是指向当前 JSP 页面本身，有点像类中的 this 指针，它是 java. lang. Object 类的实例。其方法说明如下：

(1) class getClass 返回此 Object 的类。

(2) int hashCode() 返回此 Object 的 hash 码。

(3) boolean equals(Object obj) 判断此 Object 是否与指定的 Object 对象相等。

(4) void copy(Object obj) 把此 Object 复制到指定的 Object 对象中。

(5) Object clone() 克隆此 Object 对象。

(6) String toString() 把此 Object 对象转换成 String 类的对象。

(7) void notify() 唤醒一个等待的线程。

(8) void notifyAll() 唤醒所有等待的线程。

(9) void wait(int timeout) 使一个线程处于等待直到 timeout 结束或被唤醒。

(10) void wait() 使一个线程处于等待直到被唤醒。

(11) void enterMonitor() 对 Object 加锁。

(12) void exitMonitor() 对 Object 开锁。

6. application 对象

application 对象实现了用户间数据的共享,可存放全局变量。它开始于服务器的启动,直到服务器的关闭,在此期间,此对象将一直存在;这样在用户的前后连接或不同用户之间的连接中,可以对此对象的同一属性进行操作;在任何地方对此对象属性的操作,都将影响到其他用户对此的访问。服务器的启动和关闭决定了 application 对象的生命。它是 ServletContext 类的实例。其方法说明如下:

(1) Object getAttribute(String name) 返回给定名的属性。

(2) Enumeration getAttributeNames() 返回所有可用属性名的枚举。

(3) void setAttribute(String name,Object obj) 设定属性的属性值。

(4) void removeAttribute(String name) 删除一个属性及其属性值。

(5) String getServerInfo() 返回 JSP(SERVLET)引擎名及版本号。

(6) String getRealPath(String path) 返回一虚拟路径的真实路径。

(7) ServletContext getContext (String uripath) 返回指定 WebApplication 的 application 对象。

(8) int getMajorVersion() 返回服务器支持的 Servlet API 的最大版本号。

(9) int getMinorVersion() 返回服务器支持的 Servlet API 的最小版本号。

(10) String getMimeType(String file) 返回指定文件的 MIME 类型。

(11) URL getResource(String path) 返回指定资源(文件及目录)的 URL 路径。

(12) InputStream getResourceAsStream(String path) 返回指定资源的输入流。

(13) RequestDispatcher getRequestDispatcher (String uripath) 返回指定资源的 RequestDispatcher 对象。

(14) Servlet getServlet(String name) 返回指定名的 Servlet。

(15) Enumeration getServlets() 返回所有 Servlet 的枚举。

(16) Enumeration getServletNames() 返回所有 Servlet 名的枚举。

(17) void log(String msg) 把指定消息写入 Servlet 的日志文件。

(18) void log(Exception exception,String msg) 把指定异常的栈轨迹及错误消息写入

Servlet 的日志文件。

(19) void log(String msg,Throwable throwable) 把栈轨迹及给出的 Throwable 异常的说明信息写入 Servlet 的日志文件。

7. exception 对象

exception 对象是一个例外对象,当一个页面在运行过程中发生了例外,就产生这个对象。如果一个 JSP 页面要应用此对象,就必须把 isErrorPage 设为 true,否则无法编译。它实际上是 java. lang. Throwable 的对象。其方法说明如下:

(1) String getMessage() 返回描述异常的消息。

(2) String toString() 返回关于异常的简短描述消息。

(3) void printStackTrace() 显示异常及其栈轨迹。

(4) Throwable Fill StackTrace() 重写异常的执行栈轨迹。

8. pageContext 对象

pageContext 对象提供了对 JSP 页面内所有的对象及名字空间的访问,也就是说,它可以访问到本页所在的 session,也可以取本页面所在的 application 的某一属性值,它相当于页面中所有功能的集大成者,它的本类名也叫 pageContext。其方法说明如下:

(1) JspWriter getOut() 返回当前客户端响应被使用的 JspWriter 流(out)。

(2) HttpSession getSession() 返回当前页中的 HttpSession 对象(session)。

(3) Object getPage() 返回当前页的 Object 对象(page)。

(4) ServletRequest getRequest() 返回当前页的 ServletRequest 对象(request)。

(5) ServletResponse getResponse() 返回当前页的 ServletResponse 对象(response)。

(6) Exception getException() 返回当前页的 Exception 对象(exception)。

(7) ServletConfig getServletConfig() 返回当前页的 ServletConfig 对象(config)。

(8) ServletContext getServletContext() 返回当前页的 ServletContext 对象(application)。

(9) void setAttribute(String name,Object attribute) 设置属性及属性值。

(10) void setAttribute(String name,Object obj,int scope) 在指定范围内设置属性及属性值。

(11) public Object getAttribute(String name) 取属性的值。

(12) Object getAttribute(String name,int scope) 在指定范围内取属性的值。

(13) public Object findAttribute(String name) 寻找一属性,返回其属性值或 NULL。

(14) void removeAttribute(String name) 删除某属性。

(15) void removeAttribute(String name,int scope) 在指定范围删除某属性。

(16) int getAttributeScope(String name) 返回某属性的作用范围。

(17) Enumeration getAttributeNamesInScope(int scope) 返回指定范围内可用的属性名枚举。

(18) void release() 释放 pageContext 所占用的资源。

(19) void forward(String relativeUrlPath) 使当前页面重导到另一页面。

(20) void include(String relativeUrlPath) 在当前位置包含另一文件。

9. config 对象

config 对象是在一个 Servlet 初始化时，JSP 引擎向它传递信息用的，此信息包括 Servlet 初始化时所要用到的参数（通过属性名和属性值构成）以及服务器的有关信息（通过传递一个 ServletContext 对象）。将其方法说明如下：

(1) ServletContext getServletContext() 返回含有服务器相关信息的 ServletContext 对象。

(2) String getInitParameter(String name) 返回初始化参数的值。

(3) Enumeration getInitParameterNames() 返回 Servlet 初始化所需的所有参数的枚举。

教 学 资 源 支 持

敬爱的教师：

感谢您一直以来对清华版计算机教材的支持和爱护。为了配合本课程的教学需要，本教材配有配套的电子教案（素材），有需求的教师请到清华大学出版社主页（http://www.tup.com.cn）上查询和下载，也可以拨打电话或发送电子邮件咨询。

如果您在使用本教材的过程中遇到了什么问题，或者有相关教材出版计划，也请您发邮件告诉我们，以便我们更好地为您服务。

我们的联系方式：

地　　址：北京海淀区双清路学研大厦 A 座 707

邮　　编：100084

电　　话：010－62770175－4604

课件下载：http：//www.tup.com.cn

电子邮件：weijj@tup.tsinghua.edu.cn

作者交流论坛：http://itbook.kuaizhan.com/

教师交流 QQ 群：136490705　　微信号：itbook8　　QQ：883604

（申请加入时，请写明您的学校名称和姓名）

用微信扫一扫右边的二维码，即可关注计算机教材公众号。